THE
VANISHING
OF A
SPECIES?

THE
VANISHING
OF A
SPECIES?

WHERE WE WERE

WHERE WE ARE

WHERE WE MIGHT GO

A Look at Modern Man's Predicament by a Geologist

PETER GRETENER

qⵏP

qualitas

Qualitas Publishing

A Qualitas Book

Published by
Qualitas Publishing
195 Cardiff Drive N.W.
Calgary, Alberta, Canada
T2K 1S1
www.qualitaspublishing.com

Printed on paper that is acid free, lignin free
and meets all ANSI standards for archival quality paper.

Library and Archives Canada Cataloguing in Publication

Gretener, P. E. (Peter E.)
The vanishing of a species?: a look at modern man's
predicament by a geologist / Peter Gretener.

Edited by Nick Gretener.
Includes bibliographical references and index.
ISBN 978-1-897093-82-5

1. Human ecology. 2. Human evolution. 3. Nature--Effect of
human beings on. 4. Environmental degradation. 5. Technology-
-Social aspects. 6. Civilization, Modern--21st century.
I. Gretener, Nick, 1958- II. Title.

GN281.G734 2010   304.2   C2009-904982-1

Printed in the United States of America

FOR VRENI

CONTENTS

viii   CONTENTS

ILLUSTRATIONS

FIGURES

FOREWARD

This is the book that wasn't. In my formative years, I recall my father talking often of the book he might write—*The Vanishing of a Species?*—chronicling where mankind was headed if it did not make some serious adjustments. In a nutshell, he spoke of man's infinite demands being in conflict with a finite ecosystem—planet Earth.

Years went by, and while Dad sponsored many forums and lectures on the topic, the *Book* was always "...coming, it's coming."

After a while, talk of the Book waned, save for the occasional jab—"Looks like the Book is vanishing, along with the species."

And so, colleagues, friends, family—we all resigned ourselves to the realization that the Book was something we would never experience.

Dad died on May 16, 2008.

Some weeks later we were engaged in the melancholy task of sorting through his papers. For an academic who decried the "publish or perish" mentality, he had been a prodigious author in his field of geology and geophysics. Beyond geological work, we also uncovered the odd reports, letters, and essays on topics of more universal import—along the lines of his *Vanishing* concerns. It brought back memories of the big project.

The papers were sorted out. Some were distributed to old colleagues, some were sent to the university for whatever assistance they might be to the next generation, and some—that we didn't know what to do with but just couldn't throw out—were tossed in a few boxes for storage in the basement.

Then, many months later, a call from my mother: "I've been going through Dad's boxes, and I found something you should see."

And there it was, in an old university binder, 43 chapters of the working manuscript of *The Vanishing of a Species?* It appeared to have been placed on a shelf in the late 1970s and languished there ever since.

As I turned the pages, it struck me that much of what had been put down some 30 years ago rings as true today as when it was written, perhaps even more so given the current economic turmoil. And so, while *Vanishing* almost died on the editing table—it lives.

In reading the admittedly rough manuscript, I had a choice to make. Let the words speak coarse and unadorned, as they had been penned so many years ago, or invoke the editing process—polish and smooth—to make for a softer, more "acceptable" delivery. The urge to apply my editing touches was strong, but I realized that in doing so I risked turning *Vanishing* into my work, or at best a bastardized version of what the author was trying to say.

So I left his work essentially untouched. Where I felt it necessary to add my two cents worth, this appears in the form of an editorial note rather than a modification to (subversion of?) the original text.

At its core, *Vanishing* is a plea from an earth scientist (who loved this planet both professionally and personally) for mankind, particularly the affluent-developed nations, to throttle back—to assess our real needs and adjust our expectations—in order to offer our species a chance at a sustainable future.

The author's three commandments for his *human revolution* are surprisingly simple yet vital to our survival: (i) use our greatest gift, the human brain, to its fullest potential; (ii) treat each other fairly; and (iii) limit our demands to what is needed for a reasonably comfortable, but not overly-indulgent, existence (with greater emphasis on mental and spiritual, as opposed to material, pursuits) during our fleeting time (geologically speaking) on spaceship Earth.

That *Vanishing* is not an outdated work is brought home by events all around us, every day, including the 2008-2009 global financial turbulence.

The release in 2009 of the report *Prosperity without growth? The transition to a sustainable economy* by the U.K. Sustainable Development Commission, reflects the continuing relevance, and increasing urgency, of the concepts raised in *Vanishing*:

Every society clings to a myth by which it lives. Ours is the myth of economic growth...

...It's totally at odds with our scientific knowledge of the finite resource base and the fragile ecology on which we depend for survival...

...The default assumption is that—financial crises aside—growth will continue indefinitely. Not just for the poorest countries, where a better quality of life is undeniably needed, but even for the richest nations where the cornucopia of material wealth adds little to happiness and is beginning to threaten the foundations of our wellbeing...

...Questioning growth is deemed to be the act of lunatics, idealists and revolutionaries.

But question it we must. The myth of growth has failed us...

Today we find ourselves faced with the imminent end of the era of cheap oil, the prospect (beyond the recent bubble) of steadily rising commodity prices, the degradation of forests, lakes and soils, conflicts over land use, water quality, fishing rights and the momentous challenge of stabilizing concentrations of carbon in the global atmosphere. And we face these tasks with an economy that is fundamentally broken, in desperate need of renewal.

In these circumstances, a return to business as usual is not an option. Prosperity for the few founded on ecological destruction and persistent social injustice is no foundation for a civilised society...

Prosperity consists in our ability to flourish as human beings—within the ecological limits of a finite planet. The challenge for our society is to create the conditions under which this is possible. It is the most urgent task of our times.[1]

When turning the pages, the reader should keep the following in mind:

1. Much of *Vanishing* was written in the mid to late 1970s (population of North America—Canada/U.S.— was about 250 million and world population was around 4 billion).

2. All drawings are original renderings by the author— a geologist by trade—and should be appraised in that light. Again, editing/polishing would not do them justice.

3. Organizing this work was a challenge, as all issues are ultimately intertwined (the author deplored the fragmentation of knowledge). To ease the reader's journey, generous chapter cross-referencing is employed.

4. The work is definitely not "PC"—politically correct— but it reflects the author's honest views as shaped by his times.

5. The author estimated the *crunch* point—where something will give if we do not change our ways—at around the middle of the 21st century (i.e. uncomfortably soon). While we have yet to crash in the jarring way we will if nature must correct our errant ways (as opposed to a self-imposed solution - see Figure 14.2), the clock is running down. We live in volatile times. The economy is exceedingly nervous—the future far from assured.

6. As you will see from his introduction, the author knew he was likely off-base on some of what he

wrote. He acknowledged that no one is right about everything, particularly when addressing such a wide range of issues. What is important is to think about the seminal issues and, perchance, to act. Apathy is not an acceptable option. In his own words: "Maybe none of us can point to any actual achievement. Maybe those who say that nothing can be done are right. But at least one should be able to say: 'I kept kicking.'"

I hope that *Vanishing* stirs the reader and sparks passionate debate about exactly what we are doing on this planet and where we are headed if we keep doing it. With a bit of luck and a lot goodwill—the audacity of hope? (see page 214)—those sparks just might ignite the human revolution.

Nick Gretener
Calgary, Alberta

1

INTRODUCTION

To many, this book will make peculiar reading. To cover a spectrum from man's early beginnings to the modern problems of population increase, resource depletion, pollution, crime, and many more, may not seem very sensible at first glance. However, man's ancient goal, to which he at least gives lip service and with respect to which he has made essentially no progress—the question: "What is man? Who am I?"—this question cannot be studied without putting ourselves in the larger context, both in space and time. One fatal mistake of the humanities is their almost total neglect of the advances of the natural sciences. Man is a natural phenomenon, and he cannot be understood by philosophical investigations that consider him special and essentially removed from the rest of the world, in particular all other forms of terrestrial life.

Today, we have the fragmentation of knowledge (the two cultures of C.P. Snow[1]) into humanities-social sciences and natural sciences, and more recently into many sub-fields within these larger groupings. It is this division of knowledge that prevents us from seeing the forest for the trees. The systematic downgrading of the generalists in the western world has an absolutely devastating effect. Our best brains are only responsible for a tiny fraction of human know-how. Outside interests are not required to progress on the road to fame. The result—humanists consider natural scientists, and in particular engineers, uneducated, while the latter look upon the humanists as technological and scientific illiterates. No wonder we don't understand man, who is both the most conscious form of life, with the most sophisticated social systems, and also the super toolmaker, far exceeding the capabilities of any other form of terrestrial life.

My background is that of a natural scientist or, more specifically, that of an earth scientist. Such a background is bound to colour one's outlook. But at the same time, all of us live in the human community. We deal with (have to) our fellow man in a given social system and we experience both his kindness and his depravity. So even if we are not social scientists by profession, we are part of the social interactions of man and cannot help but develop some expertise in that field in the course of our lifetimes. The humanities deal with those aspects specific to man and man alone. In particular, humanists reflect on man's past, present, and future. And again, most of us have done so from time to time and have something to offer, even if it may not be as well thought out as from those who make it their profession.

It is obvious from the aforementioned that I deplore the fragmentation of knowledge. It is equally obvious that I reject the cherished North American idea that one needs a union card (read PhD) in a given field in order to voice a valid opinion. I would agree that one must do one's homework and must be prepared to substantiate one's reasoning. But I do not buy the concept that each little field of human endeavour represents a closed shop affair to which outsiders have nothing to contribute. I believe the examples are numerous where outsiders (read mavericks), or people without formal training in a given field, have made outstanding contributions.

It is for these reasons that the thoughts presented here are wide-ranging and not restricted to a partial view of man. The risk is evident. One must read all that is said with a critical mind. No single person can be right on everything, even though some do believe in this concept (e.g. Mao). In this context, it is worthwhile considering an excerpt from G.K. Gilbert's presidential address to the Geological Society of Washington, D.C. back in 1895:

> The method of hypotheses, and that method is the method of science, founds its explanations of Nature wholly on observed facts, and its results are ever subject to the limitations imposed by imperfect observation. However grand, however widely accepted, however useful its conclusion, none is so sure that it cannot be called in question by a newly discovered fact. In the domain of the world's knowledge there is no infallibility.[2]

The following words represent the ideas of a single contemporary with a look into the past as best we understand it today, a look (naturally biased) at the present, and a glimpse at the possibilities of the future.

YESTERDAY

2

MAN AND SPACE

Whenever in the dim past man's level of consciousness reached a sufficient level, he must have looked up at the firmament and felt an uncomfortable sense of humility. Not for long, though. Soon he realized that the heavens seemed to stretch to infinity in all directions, and he was quick to reach the conclusion that he must be at the centre of all things. This in no little way restored his confidence in himself. It confirmed his belief that the emergence of man was the long-awaited event and that the universe was created for his express purpose.

When, almost 400 years ago, Galileo had the audacity to suggest that the earth circles the sun rather than the other way around, this turned out to be a most unhealthy endeavour. Not that he was a liar. This, by any human standard, would only be a minor offence. On the contrary, he was right of course, as we all know, and thereby got in conflict with the human ego. The belief was then, as now, that man represents the ultimate in creation, and his emergence is the sole purpose of the universe. This naturally demands that he be at its very centre.

But worse was yet to come. The solar system was recognized

as a star with a series of planets. The earth being neither the closest nor the farthest. It is, however, in a special position in so far as it maintains just the right distance from the sun to have a temperature regime which permits the existence of life as we are used to it. Small consolation.

Additional investigation showed that stars with their associated planets occur in galaxies. In our own galaxy, the Milky Way, our sun occupies a rather peripheral position. Further disappointment.

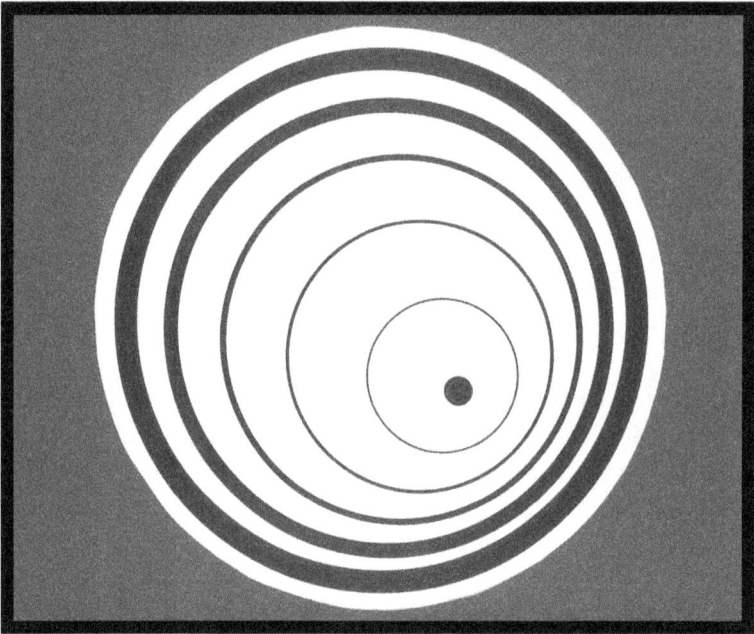

INFINITY = ETERNITY

The universe, as such, contains many galaxies, and we now realize that our Milky Way is in no way at the centre of all happenings. In fact, accepting certain theories, there may well be nothing at the centre at all.

Where does all this leave us, the unique species of *Homo sapiens*? Well, one thing is abundantly clear: *We* are way off centre, and in more ways than one, as we shall see.

But when discussing space we are not only concerned with where we are in space, but also whether or not we are *the only* ones. This is an ancient question and has occupied man's mind throughout his history. Of course, at this moment there is no hard proof of the existence of extraterrestrial life, despite all of the UFO stories. However, simple statistics tell us that we are not likely to be the only ones.

It is true that the probability of finding a star-planet combination such as the sun-earth, that provides the proper conditions for the development of organic life as we know it, is low. However, the number of solar systems in the universe is so immense that the probability for occurrence of life on at least a few planets must be considered almost a certainty. We must further be aware of the fact that the evolution of life on earth is very much a haphazard process that has been going on over several billion years (see Chapter 3). It is therefore most likely that somewhere in the universe this process has produced life that is further advanced than our own.

Thus, the consideration of the space around us does little to enhance our ego. In fact, it introduces an element of humility to our thinking.

3

MAN AND TIME

Today, we are well acquainted with short time spans. Cameras are taking pictures of bullets flying at incredible speeds, rockets bridge the gaps in space travelling at undreamed of velocities, computers produce instant answers, parents drink instant coffee, and kids have instant wisdom.

Well, in this chapter we shall take time out from our fast living world and contemplate the very opposite, very long time spans, or what often and commonly would be referred to as "eternity." It is quite obvious that we all (including the specialists—the geologists and the astronauts) have a poor sense of appreciation for such matters as the age of the earth, ancient life, and early man or even what by geological standards would be called *modern* or *recent* man. Whether we are fully conscious of it or not, our yardstick for long time spans is our lifetime—somewhere in the range of 50 to 100 years. But what a poor yardstick to try and come to grips with millions or even billions of years. Evidently, this seems an almost hopeless endeavour.

So let us start by taking a journey into the past, a brief one at that, and explore the rise of our species and our present position against the background of the earth's history. One thing will become painfully evident very quickly. The earth has done extremely well for a very long time without our distinguished presence—a blow to the human ego, which today is as vulnerable as in Galileo's time.

To look into the face of time, let us proceed in the following manner. Geologists have found that the oldest rocks on earth are in the order of 3,500 million years old. This is not a meaningful number. Let us therefore assume that the oldest rocks are exactly 3,650 million years old (a couple of hundred million years are of no significance to the spirit of the argument we want to present). We now squeeze this rather ludicrous number into ONE year by assuming that each day represents 10 million years. See Figure 3.1 for a presentation of the rise of man in geological time.

The oldest rocks were formed on January 1st of our year, while right now we are celebrating New Year's Eve, i.e. December 31st, 24:00:00. The need for the highly precise definition of "right now" will become apparent as we proceed. The earliest traces of life are found in rocks 3,200 million years old,[1] which place them into mid-February of our year. Abundant higher life with hard shells and skeletons that fossilize well only came into existence some 600 million years ago, or about November 1st. No trace of man yet.

The land was conquered first by plants and a little later by animals around mid-November. The mammals, of which we are a part, appear at Christmas. But still no trace of man. Forms which deserve the title pre-man (any self-respecting *Homo sapiens* would deny their kinship) make their appearance on December 30th. If we accept the definition *man the toolmaker*, and this seems reasonable (see Chapter 5), the oldest remains found so far date back about three million years. Thus, the toolmaker comes into existence on December 31st at about 17:00. Modern man, the one who left us the beautiful cave drawings in southern Europe and North Africa, dates back as far as 30,000 years or December 31st, 23:55:00. The agricultural revolution is placed at about 10,000 years ago or December 31st, 23:58:20. And the industrial revolution began some 200 years ago or December 31st, 23:59:58.

FIGURE 3.1
MAN AND GEOLOGICAL TIME

Event	Actual Time	Time Reduced to One Year (Rough Scale*)
Homo Technicus	200 Years	Dec. 31 23:59:58
Agricultural Revolution	10,000 Years	Dec. 31 23:58:20
Homo Sapiens	30,000 Years+	Dec. 31 23:55
Homo Erectus	500,000 Years+	Dec. 31 23:00
Earliest Hominid (Tools)	3,000,000 Years	Dec. 31 17:00
Departed From Ape Line	20,000,000 Years	Dec. 30 00:00
Rise of the Mammals	70,000,000 Years	Dec. 24
Land Conquered	400,000,000 Years	Nov. 21
Abundant Fossils	600,000,000 Years	Nov. 1
Oldest Fossils	3,200,000,000 Years	Feb. 15
Oldest Dated Rocks	3,650,000,000 Years	Jan. 1

* One Day   = 10 million years
  One Hour   = 400,000 years
  One Minute = 7,000 years
  One Second = 100 years

The message is clear—man is a newcomer. He has not been around for a long, long time as common belief would have it. He evolved relatively recently. He is an integral part of our planet, but not an essential one. There is no natural law that guarantees his survival. Just like all other forms of life, his evolution is an ongoing process, making it likely that our successors five million years down the road will be as different from us as we are from Australopithecus. The important thing is to see to it that there *are* such successors.

4

FURTHER IMPLICATIONS OF LONG TIME SPANS: AWARENESS OF EVOLUTION AND THE RARE EVENT

The Present is the *key* to the Past (emphasis mine) reflects the arrogant attitude of the newcomer, *Homo sapiens*. Nobody will dispute the fact that the study of the present can reveal many important aspects of the past. But the above proverbial saying at least implies that *everything* about the past can be learned from looking at the world around us. This is certainly not the case for several reasons, two of which are touched on below.

First is the fact that many factors have been changing gradually—thus, there was no significant terrestrial life for about 90 percent of total earth history (see Figure 3.1 – land is conquered on November 21st of our one year earth clock); terrestrial heat flow has decreased by a sizable factor over the life of the earth; the composition of the atmosphere has undergone drastic changes since the dawn of earth history; and many more.

Second, the present provides no, or only vague, indications of the infrequent events that must have occurred in the past.

The fossil record gives clear evidence that "unusual" conditions must have prevailed at certain times during the last 600 million years. Events of such low frequency are not amenable to rigid study. They remain in the realm of speculation, since it is only their effects that can be detected in the fossil record.

In view of the two objections stated, it seems more realistic to say, "The study of the Present offers a keyhole view of the Past." The current discussion centres primarily on the *angle of vision* provided by this keyhole.

When considering the long life of the earth, we become aware of two things. First, the change in or rise of life, commonly referred to as evolution. The fossil record speaks clearly. Early rocks bear traces of primitive life as far back as 3,200 million years ago. Younger rocks contain remains of increasingly more diversified and complex life. At the present time, man is the culmination of this process. No such advanced toolmaker has lived before us; otherwise the rocks would bear witness to such activity as we see it today.

The fact that in many cases we can observe a slow and seemingly continuous change in the forms of life in adjacent rock strata suggests we have no reason to invoke continuing creation. The processes of mutation and natural selection which are operative today, together with Father Time, are perfectly sufficient to account for such change. This does not resolve the question of the origin of life, which despite some hypotheses must still be considered a mystery, though it may be elucidated with time.

For those who do believe that essentially continuous creation is called for: Why do you believe that the Lord should preoccupy himself with us in this manner? If you believe in the existence of God, you certainly must acknowledge that he has to watch over the whole universe. To imagine that he would spend all of his time on us, to the extent of helping to win the weekend football game, just defies all logic. Be that as it may, it is fair to say that the process of biological evolution is now reasonably well understood and supported by the evidence found in rocks.

Just what are the implications? One, of course, is the fact

that evolution is an ongoing process, which possibly may be directed but certainly cannot be stopped. As Simpson says, "By no standard of ethics whatever is human society now so good that any ethical man could wish it to persist unchanged or could fail to hope and to work for its improvement."[1] This puts it quite squarely that not only can change not be stopped, it is not desirable to stop it. However, it is desirable to *direct* the change and this possibility is almost within our grasp, as evolution and processes sustaining it become better and better known.

But before the change can be directed, the question must be decided as to where we should go. As long as there is no agreement on this matter, genetic engineering is not only premature, it is criminal. To quote Simpson again: "The probability of survival of individual, or groups of, living things increases with the degree with which they harmoniously adjust themselves to each other and their environment."[2] It is hard to argue with this statement as a natural scientist, and at the same time it is equally clear that the above description does *not* fit man as we presently know him.

It is clear that we must leave this at present an unsolved problem, but one worthy of attention. No other form of life ever had this weapon of knowledge when threatened by extinction. We do have it, but time is running out since we have concentrated all of our efforts on tool making. The result may well be a short period of extreme dominance followed by an equally rapid disappearance—a *flash in the pan*. In fact, one may say that this has to be so since it is the only way nature can preserve itself and guarantee the continuation of life on earth. A predominance, as we aspire to it at present, is not tolerable on a worldwide basis for any length of time as it would threaten the very existence of life as such.

Studying the rise of life through geological time leads us to a second concept that should concern us here—that of the *rare event*.[3] Events of low probability are classified as impossible on the human time scale yet figure as certainties on the geological time scale.

It is important to distinguish between the *improbable* and

the *impossible*. I postulate the improbable when I claim that a major meteorite will hit the Atlantic within the next five years. I am not likely to be right, but there is no physical law that would preclude such an event. For the tabloid "News" to declare "Baby Born with Wooden Leg" is to advocate the impossible, since wood is not known to grow in the womb. An event with 95% probability of occurrence in the next 15 million years is not a major concern to the human race. However, to relegate it into the realm of the impossible is not permissible for those studying the earth's history.

The rare event has a definite significance in geology and is defined as an event (spasm or episode characterized by a fast change) with the low probability of a particular interplay of various factors. For example, Figure 4.1 shows the probabilities associated with the *8-Dice-Game* (the number is arbitrary).[4] If we have eight dice and wish to throw eight sixes, our chances of doing so in the first throw are extremely small, about 1 in 1.7 million. If we only play dice for an hour and make, say, 200 throws, the probability of throwing eight sixes at least once will still only be about 1 in 10,000—negligibly small. In day-to-day life, probabilities of such low order tend to be called "impossibilities." Not because they are physically impossible— there is no physical law that prohibits eight sixes from turning up—but rather because the probability for one occurrence in a few trials is so low that it can be safely ignored.

Yet in earth history, the game has been going on for a long time. What is proper, or at least permissible, for one evening's game no longer holds when the *8-Dice-Club* plays night after night. Father Time is a persistent player.

If we raise the number of trials to 1.7 million, the probability of throwing eight sixes will increase to 63 percent, and for 5 million trials it will be more than 95 percent. Thus, the improbable becomes probable and eventually approaches certainty. This trivial principle has applications in geology, where life spans are long and consequently the number of trials is large.

In addition to truly rare events, one can also consider the timescale as it applies to such phenomena as major storms,

FIGURE 4.1
PROBABILITIES ASSOCIATED WITH THE 8-DICE GAME

Results	Basic Probability	Throws Required for 95% Probability of at least One Occurrence
8 Sixes	1 : 1,700,000	5,000,000
7 Sixes	1 : 42,000	120,000
6 Sixes	1 : 2,400	7,000
5 Sixes	1 : 200	700
4 Sixes	1 : 30	100
3 Sixes	1 : 10	30
2 Sixes	1 : 4	12
1 Six	1 : 3	8
0 Sixes	1 : 5	13

flash floods, landslides, and others. Clearly, these are rare only by human standards. From a geological point of view, such phenomena are regular episodic agents. In terms of the *8-Dice-Game*, one can speculate that the occurrence of seven sixes will also produce an event, even though of less dramatic nature than eight sixes. Equally, six sixes will not go unnoticed, and so on. As the effects become less shattering, the occurrence will be more frequent. Figure 4.1 provides the basic probabilities involved. From Figure 4.1 follows another triviality—*big events are rare, small events are common*.

An attempt to classify events according to their rate of occurrence appears in Figure 4.2.[5] The rationale behind the arbitrary selection in Figure 4.2 is purely pragmatic. *Regular Events* figure in human life (e.g. the 100 year flood of the engineer). *Common Events* must appear in the recorded human history, even though they usually are no longer perceived as the truth, but rather ascribed to the vivid imagination of the ancient recorder. *Recurrent Events* are important in terms of the fossil stage. *Occasional Events* are the type responsible for the major evolutionary breaks and *Rare Events* are truly just that, having occurred at most very few times through earth history.

As an aside, no discussion of the implications of long time spans is complete without mentioning the *haze of the past*— loss of resolution with distance in time and space—a well known phenomenon, to say the least. Yet it is often forgotten. Every human historian must reckon with the fact that things get more hazy and fuzzy as one moves towards "the dawn of civilization." This, of course, is even truer for the geologist, who deals with time spans almost a million times longer. To forget the concept of the *haze of the past* inevitably leads to false conclusions when the present is compared with the past.

It has, for example, been suggested that magnetic field reversals are more frequent in the recent than the distant past, a suggestion that implies the magnetic field is getting more "nervous." One may well ask: "Is this true, or is it simply an apparent effect due to our sharper perception of the recent past?"

FIGURE 4.2
ATTEMPT TO CLASSIFY EVENTS
ACCORDING TO THEIR RATE OF OCCURRENCE

Type of Event	Number of Years Required for 95% Probability of at least One Occurrence
Regular Event	100
Common Event	1,000
Recurrent Event	1,000,000
Occasional Event	100,000,000
Rare Event	1,000,000,000

THE HAZE OF THE PAST

Returning to the rare event, the origin of life, as noted in the previous discussion on evolution, is by no means a settled question—there are many different schools of thought. At least one authority in the field, G. Wald, believes that the rare event was instrumental in the formation of life. Wald interprets the origin of life as a series of spontaneous events. Each event has a low probability, but there is no way of assessing this probability. In the case of the formation of life one is, according to Wald, not even able to define clearly what constitutes a trial. However, as time passes, the number of trials, whatever their nature, is bound to increase.

Wald discusses in an abbreviated fashion the mathematics of the problem and concludes his statistical treatment with the following observation:

> Time is in fact the hero of the plot. The time with which we have to deal is of the order of two bil-

lion years. What we regard as impossible on the basis of human experience is meaningless here. Given so much time, the 'impossible' becomes possible, the possible probable, the probable virtually certain. One has only to wait: time itself performs the miracles.[6]

Needless to say, Wald's "miracle" implies nothing supernatural. It is merely a natural process operating completely within known physical laws but of such a low probability that its occurrence strikes our human minds as a miracle. Wald's theory opens up the possibility that the rare event is of prime importance in the history of the earth and its inhabitants.

The fossil record quite convincingly demonstrates that evolution does not proceed at a more or less constant rate. There have been moments in the past when a very rapid change in the composition of life took place. Rapid is used in the geological vernacular and may well mean several million years. Yet it is clear that at certain times various life forms that felt comfortable on the earth for many millions of years suddenly disappeared. Such *crises in life,* as Newell[7] called them, have undoubtedly taken place in the past, and again the record speaks clearly.

Of course, the rocks only preserve the consequences— mass extinction—and not the event leading to such a drastic change. As a result, we are left to speculate. The fact that major breaks in evolution are rare and occur on the average maybe every 100 million years, does not make our task easier. It simply points to the fact that whatever the reasons, it must be a very unusual set of circumstances.

In terms of faunal breaks (extinction), the *8-Dice-Game* may take the form suggested in Figure 4.3. It is not to be taken at face value, but rather should be viewed as an illustration, and it is not implied that all the factors listed have to coincide in order to produce a major evolutionary break. It is far more likely that the number of decisive factors is smaller, but their probability of occurrence is much less than one-in-six, as in the case of a simple game of dice.

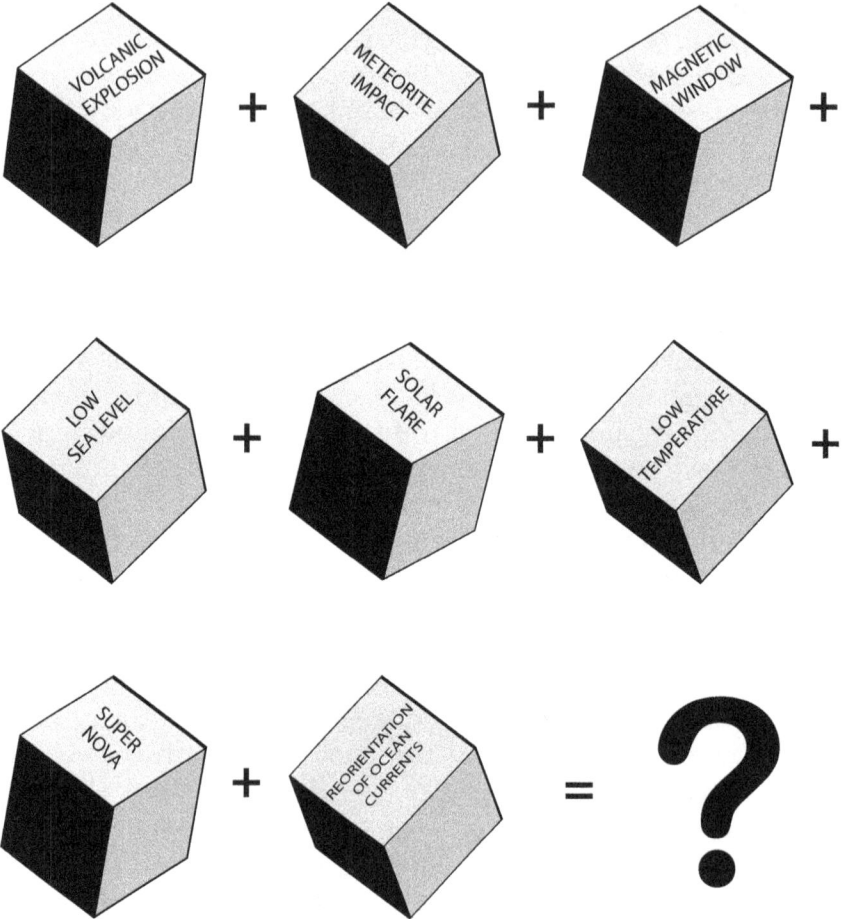

FIGURE 4.3: THE RARE EVENT OR THE 8-DICE GAME.
This is an illustration not a model. Listed on the dice are
various factors which have been advanced as responsible
for abrupt faunal breaks (extinction). Coincidence of several
or all such factors will be rare and will produce one of these
elusive breaks.

Since man's existence on this planet—his oral and written record—does not encompass more than 10,000 years, evidently such catastrophic events do not figure in our direct observations of the world around us. That does not mean they are impossible, merely improbable, and there is no natural law that would preclude their reoccurrence within the next minute, the next decade, or the next century.

As an example, we may take the collision of the earth with a truly major meteorite. There is no longer any doubt that cosmic bodies of various sizes have hit the earth numerous times throughout its history. Although the impact of a very large meteorite is unquestionably a rare event, it is nonetheless intriguing to speculate with Dietz on the far-reaching consequences of such an occurrence.[8] A fall into the sea could give rise to tsunamis of unheard-of size, which might well wipe out many coastal areas where human population is currently concentrated. Impact on land would produce an earthquake of high magnitude, possibly in an otherwise aseismic area. The fall of a large meteorite is a singular event with both a low probability of occurrence and a low probability of survival. When an air crash with 100 dead makes headlines around the world, it staggers the imagination to think of an event that might wipe out tens of millions. However, we are not in a position to guide our destiny in this regard. Life has always been risky and will remain so. The joy of being born includes the inevitable acceptance of death.

The possibility is quite real that the events leading to the major breaks in evolution had, in fact, different natures. Whether we survive or not, one thing is sure—our presence has already caused such a break. The extinction of species due to our presence is estimated at several hundred,[9] and we are not just concerned with the disappearance of the Peregrine Falcon, as Beckerman[10] suggests. It can, therefore, not be ruled out that we may be the cause of our own extinction. While we cannot control the natural rare event, we should try to avoid creating one ourselves.

Simpson states: "Man was certainly not the goal of evolution, which evidently had no goal."[11] Teilhard de Chardin says,

"Evolution = Rise of consciousness."[12] Does this represent a contradiction? Not at all. Simpson refers to the fact that evolution proceeds by trial and error and is affected by outside parameters, such as changing climatic conditions and others. To predict the course of evolution in a precise manner is, therefore, a hopeless endeavour, and the fact that specifically man now represents the highest form of life on earth must be termed an accident. De Chardin, on the other hand, refers to the clearly observable fact that throughout time life becomes more diversified, more complex, and more conscious. This general trend is followed by the circuitous path of trial and error.

LIFE IS ALWAYS RISKY

Better definition of evolution: Evolution is the rise of complexity, which in turn leads to both diversity and consciousness. The first two are directly observable in the fossil record, the third is not. In fact, while we can explore the level of consciousness in man, we are still very much at a loss as to how to assess the presence and level of this property in other forms of terrestrial life. To presume that consciousness at any level is the prerogative of man alone seems premature at this stage. That his level of consciousness exceeds that of animals by orders of magnitude is presumably correct.

Re Simpson, the fact that evolution proceeds by trial and error makes it clear that while increased consciousness must evolve, the human consciousness represents only one form and alternate developments are certainly possible. In this regard, it is wrong to consider human consciousness as either the ultimate or the only form of consciousness. Since it took over three billion years from the inception of life to reach the human level, it is equally clear that evolution proceeding on a similar track on another planet could by this time have produced a superior level of consciousness (see Chapter 2).

The study of biological evolution lies within the sphere of the paleontologist while the study of cultural evolution is the field of the archaeologist and anthropologist.

5

FROM EARLY MAN THE TOOLMAKER
TO PRESENT HOMO TECHNICUS

For us as outsiders, it is always of interest to know at what stage an anthropologist or geologist should call a fossil remain that of man or designate it as *Homo.* Leakey called his find in the Odulvai gorge *Homo habilis,* on the basis of stone tools that were found adjacent to his remains.[1] This find was dated as almost two million years old. More recent findings indicate that this date for *man the toolmaker* may eventually be pushed back as far as five million years. This is not because the science is underdeveloped, but rather because unravelling man's past is a most frustrating job.

Man is a land dweller, and the mortal remains of all such life preserve very poorly. Obviously, such skeletons are subject to transportation and destruction by running water, wind, frost, and other agents after deposition. As a result, the long-term history of such life forms, which is based on fossil finds, is very difficult to establish. This in contrast to forms living in the sea, where conditions for preservation range from good to excellent.

Much of the anthropological controversy that continues to exist is due to today's still very fragmentary record of man's

past. However, this is not the point we wish to address ourselves to. The question is rather: "What is man? His erect
posture, his language, his large brain, his highly developed
cultures?" Some of these traits can be deduced from the fossil
record and some cannot. Obviously, we have to pick one which
is recognizable. The erect posture, of course, can be inferred
from the nature of the skeleton, provided a find is sufficiently
complete. The same is true for the brain size, and it has been
used extensively.[2] However, to conclude that brains of certain
types were different in size on the basis of what is usually an
insufficient sample is very dangerous. Let us not forget that
presently, healthy human brains range from about 750 to
2,000 cubic centimetres.[3] As noted, Leakey chose his name
Homo habilis on the basis of associated stone tools. Later, this
was questioned since it has been found that certain animals
are also tool users. Let us briefly review this whole question.

When talking about tools, we may first of all distinguish
between *biological* and *inanimate* tools.[4] Biological tools are
claws, feet, flippers, beaks, and hands. Such tools permit
locomotion, feeding, and other necessary activities. They are
part and parcel of the organism. A particular biological tool
may provide diverse uses. For example, a paw is a tool for locomotion, but if it is equipped with claws, it becomes a digging
tool or the deadly weapon of a predator. The most outstanding, and often forgotten, biological tool is the human hand,
which is capable of the power *and* the precision grip (index
finger on thumb).[5]

Inanimate tools are objects such as sticks, rocks, and other
objects used for a certain purpose. The inanimate tools can
further be classified as:

Type 1 - Selected Tools
A natural object is selected that fits the purpose and is used
in unaltered form.

Type 2 - Selected and Modified Tools
A natural object is selected and subsequently modified before
use.

Type 3 - A Tool from a Tool
A natural object is selected and modified for the express purpose of fashioning a secondary tool from it.

The above classification is obviously one of increasing complexity, requiring increasing intelligence on the part of the user.

Leakey's definition has been challenged on the basis of new observations by animal behaviourists finding animals as tool users. However, most of these fall in the Type 1 class of selected tools.[6] The only observation to date of tool use of Type 2 (selection and modification) is by van Lawick-Goodall on chimpanzees, which have been observed to chew leaves and use the wad as a sponge, or strip leaves from twigs to fish for termites. This refers to observations in the wild. Many animals have a higher potential, as has been demonstrated by training in captivity. What does this have to do with Leakey's definition? Is it indeed outdated? Note that Leakey's tools are of Type 3: a tool from a tool. No such use has been observed in free living animals.

We may argue that the science of animal behaviour is young, and we do not yet know all there is to know. Could the tool classification system be changed? Not likely. Consider the necessary prerequisite to be a toolmaker of Type 3. You need a well developed brain in combination with an outstanding biological tool, such as the human hand. The only being that qualifies is man. A dolphin, obviously a highly intelligent animal, is never going to be a competitor—what can you achieve with flippers? Biological evolution has placed quite stringent limitations on the other forms of life. Only those with prehensile limbs qualify, and of those, man is the outstanding example.

Over the last few thousand years, man the toolmaker has: (i) developed the wheel, the boat, and gunpowder; and (ii) harnessed animal power, fossil fuels, and atomic energy. The definition of man the toolmaker is a valid one. It is this capability that has made man the most dominant species ever. It seems appropriate to associate the term *Homo* with the early origins of this particular trait. Unfortunately, an originally beneficial trait can be overdeveloped (*hypertely* of the paleontologists) to where it becomes detrimental. And this seems to be the stage which we have presently reached.

6

HOW UNIQUE IS MAN?

Some people will maintain that the question "How unique is man?" is an impossibility—that anything either *is* unique or it is *not*. I dispute this and say that the above question is not only fully justified but most relevant to our understanding of ourselves and our nature. It can, of course, not be divorced from the contemplation of our early history, our evolution, and our present relationship to other forms of life on this earth.

This is always, even in this day and age of enlightenment, a touchy question. Any negative observations are likely to rub against our ego and, therefore, will be discounted as invalid.

There can be no question that man is unique in many ways. Possibly the most drastic is his level of consciousness, which seems to exceed that of any other form of life by a considerable margin. He is also unique in that he is not only endowed with a superior brain, but at the same time he has the most advanced biological tool, the hand. It is this combination that has led to man's total physical domination of the planet.

Yet in other ways, man is definitely not unique. He is dependent on an atmosphere of the present composition, just like most other forms of land life. He is mortal—even though he finds this hard to accept—like all other forms of life. He is dependent on sources of food, and the processes of life within him are those observed in most other animals. Simpson talks of man's personal responsibility being perhaps his most unique feature, but he notes that: "Man is *part* of nature, and he is *kin* to all life" (emphasis added).[1] Thus, the conclusion is inevitable that: *Man is unique within limits.*

This is a most important realization. Devotees of man's religions, and even those of non-religious persuasions, have always tried to establish a sense of absolute uniqueness, a uniqueness that would free man totally from all the restraints placed on other forms of life on earth. This is a dangerous fallacy. It has been said that man is independent of the environment—that he carries his own environment with him. This is only true to a very limited degree. In hostile environments there are two choices. Man can survive in limited numbers under less than comfortable circumstances, or he can settle such places en masse but at a tremendous price in resources, particularly energy. Anyone living in an other than truly temperate climate today must be well aware of this fact.

While man is certainly a unique creature in many ways, he is not exempt from obeying natural laws. Such laws may be transgressed temporarily in some cases, but in the long run such disobedience spells disaster. To accept man's total uniqueness and abolish all taboos, as we have done in the 20th century, can only have devastating effects.

There is no natural law that guarantees man's continued existence.

7

MAN: A PRODUCT OF
HIS NATURE AND HIS NURTURE

The question of whether man's behaviour is dictated by instincts inherited from his animal past, or is totally a product of his environment and essentially all of his responses are learned, is strongly debated by the anthropologists and their associates in related fields. It is a question that concerns all of us. Dobzhansky[1] refers to this as *nature* and *nurture,* and these two brief and highly descriptive terms are adopted here. The particular human trait under discussion is man's aggression. Is it innate or learned as a consequence of our biological or our cultural evolution? That modern man over the last 10,000 years has been, and is, highly aggressive is not disputed by anyone. That this aggressiveness forms one of the central problems in our current predicament is equally undisputed. Thus, its cure or suppression is of foremost importance. In order to find a cure, one must know its origin. This quest to understand man's nature and its consequences for our future is pursued in Chapter 8.

One point on which there seems to be general agreement is that crowding enhances latent aggression or invokes aggression. This appears to be true for animals as well as

41

man. Scarcity of room, resources, or other vital ingredients is likely to lead to a tooth-and-claw ethics. Thus, the many fishery disputes which we have seen lately seem to indicate that the fishery fleets no longer operate in an era of plenty. This seems to confirm reports by concerned conservationists who claim we are *mining* the oceans rather than *harvesting* them. The fact that crowding has such a negative effect seems rather obvious. It is equally obvious that crowding is on the increase, even if one wishes to avoid such inflammable terms as population explosion.

In this context, it is interesting to note that all concerns about our increasing human population are directed toward feeding the additional mouths, as well as providing adequate shelter. Few, if any, authors ever contemplate that there might be a critical population level beyond which "man simply goes ape," the reverse of what we normally contemplate. Since we evidently do not love each other, to exceed a critical density in packing may be fatal from a social rather than physical point of view.

Do we have the ability to throttle back or tame our aggression to increase our odds of survival? Lewis and Towers state: "Man now abandons the rigidity of instinct for an altogether new plasticity."[2] Why does human history not corroborate this statement? Surely, as the authors state, tools have been improved, but human nature as far as we can tell has not changed over at least the past 10,000 years. Granted, earlier assessments of human nature or pre-human nature are of a more speculative nature, but high plasticity (ability to change) is not compatible with the stubborn retention of human depravity, eventually resulting in our present predicament. It is clear that the major religions preaching to give your fellow man a fair shake are correct in terms of man's survival—and yet, several thousand years of preaching have left no mark!

Montagu states: "Man's suppleness, plasticity, and most important of all, ability to profit by experience and education are unique."[3] This is exactly what the *other side* questions. Human history does not provide any indication that our ability to do these things is as outstanding as we would like to think. We do have the intelligence to recognize our faults, and we do recognize their absolutely devastating consequences, but *something* prevents us from taking action and improving our own selves. To stipulate that there is more to man than consciousness, and that there are inner forces (instincts if you wish) that interfere with our efforts, is not so far-fetched. It is the instinct—the program—that commands an animal to take a certain action in a given situation. In man's case, reason would dictate that we do one thing and yet, time and again, we do something else that contravenes reason. Those that claim that instinct-like forces must be at work in man cannot be all wrong, otherwise human destiny would be determined by human reason—and who would want to make that claim?

8

THE QUEST FOR MAN'S NATURE

As noted in Chapter 7, the quest for man's nature centres on man's aggressiveness. We should note that the term aggression refers to behaviour directed against co-specifics, i.e. man against man, tiger against tiger. It does not include the behaviour of the predator displayed against its prey.

That man today is hellishly aggressive remains an undisputed fact. All one has to do is drive a car around one of the great cities in our "highly civilized" world. The question that is under dispute is twofold:

1. Is this aggressiveness an "animal instinct" or is it a learned trait?

2. Is it an old or relatively newly acquired characteristic?

According to the main proponents, I call this the Montagu versus Ardrey-Lorenz controversy. The latter are from the outset at a disadvantage. Ardrey is a playwright who does not possess the union card of a PhD in the field. Professor Montagu is quite careful to clarify this by referring meticulously to "Mr. Ardrey," clearly expressing the view that his endeavours must be classified as unforgivable trespassing. The fact that this Mr. Ardrey has devoted many years to the study of the subject does not count, since his efforts have not been officially sanctioned by university degree. Konrad Lorenz is acknowledged as the father of ethology. The American social science whiz kids have promoted him to grandfather—somewhat outdated and certainly no longer *with-it* in our youth oriented society. So one must be clear from the outset that the Ardrey-Lorenz team is the underdog in this controversy, at least as far as North America is concerned.

In the following, it is equally clear that it is sheer madness for a mere geologist to dare offer his opinion on the matter. Let us take the above questions one by one.

First, the question of instinct versus learned. Dart/Craig,[1] and after them Ardrey[2] and Lorenz,[3] have argued that the fossil record indicates that even man's ancestors, the Australopithecines, were aggressive, and that many died from their fellows' hands. They therefore concluded that man's aggression is inherited from the animal ancestry and is in his genes. Montagu,[4] and with him most North American anthropologists, have termed this the *killer instinct* of man and they have taken violent exception to the suggestion that man should be thus endowed. In their view, this is a relatively recent phenomenon of the cultural evolution. In fact, some argue that even Pleistocene Man, some 20,000 to 50,000 years ago, was a peaceful, cooperative hunter.[5] Essentially, in the hunting stage of man's development, cooperation had such enormous survival value that the assumption of great human aggressiveness would not seem permissible. Being pragmatic outsiders, we may say that the current problem is to reduce or eliminate human aggression (current defence spending for many nations is substantial and that clearly spells doom).

From reading the literature, it is obvious that the American scientists have abandoned the word instinct, which in fact is a dirty word. Man is all learned, a total product of his environment, both pre and post natal. While our outer shell—

tall versus short, slim versus obese, etc.—is determined by our genes, these contribute virtually nothing towards our character. Man is born as a clean slate.

Now the question poses itself: "Who in this controversy are the extremists?" Nobody, including the Ardrey-Lorenz side, disputes that learning is a most decisive factor in man's life; that man's learning ability is outstanding and unique in terms of terrestrial life. But to say he is all learned makes him an end-member of a continuous process, and end-members are generally rare. Thus, to say that the only instinctive traits in man are such things as sneezing, laughing, and scratching simply does not make sense. It is also contrary to the experience every one of us accumulates as we travel through life.

Of course, one of the contentions of modern science, and in particular social science, is that all things are far more complex than they appear on the surface. The application of common sense is therefore not permitted, and from observing the social sciences it is obvious that they have successfully abolished anything remotely resembling common sense. To argue that human aggression cannot be an instinctive trait, simply because man has no instincts, is not an acceptable argument.

Following Ardrey and Lorenz, our aggression is a product of our biological evolution. Accepting Montagu, and with him the overwhelming majority of researchers, it is caused by our cultural evolution.

What difference does this really make? Is this a purely academic argument, or does it have consequences for all of us? Unfortunately, the literature always becomes vague at this point, and we are left very much alone. There is agreement that our current aggressiveness, coupled with our technical potential, is a lethal combination, which could produce a rather sudden extinction of the species. To combat aggression, we must know its roots. But to claim that one view is more optimistic than the other is really not right.

In essence, Ardrey and Lorenz see man as a victim of the conflict of his biological and his cultural evolution, while Montagu and others see him depraved by his cultural evolution alone. Just where the pragmatic difference lies is difficult to say. Adhering to the Ardrey-Lorenz view, we would have to assume that any salvation of man requires the cultural evolution to be capable of overriding the biological one. In the case of Montagu et al., a reversal in the cultural evolution is called for. Since it is generally accepted that evolution is irreversible, both views—in terms of optimism for the future—are about equally dismal.

In either case, the battle against human nature will be a tough one. However, we must not forget that for the first time in the earth's history, a form of life is aware of the process of evolution—both biological and cultural—and its implications. And not only is man aware of the process, he is also far enough advanced to guide it. It would presumably be preposterous to assume that he could stop it, nor would this be desirable at the present stage.

The case becomes even more unattractive when one considers that time considerations demand not only a cultural evolution but rather a revolution—a very abrupt turn-around (see Chapter 47). The advocates of human aggression being a learned trait accuse the instinct group of spreading hopelessness, the attitude: "I am born that way, there is nothing that can be done about it." However, those that support the view of a learned trait make it clear that its geologically recent emergence is probably related to the concept of personal property evolved during the agricultural revolution. They would thus have to advocate the abolishment of property, and, reading Beckerman,[6] one can see that this idea enjoys little popularity. The argument, forcefully as it has been carried, may thus largely be an academic one, neither side being able to offer a real ray of hope.

One cannot help but feel that the followers of Montagu desperately try to paint an uplifting picture of basic man and re-erect the *noble savage*, as stated by Ardrey.[7] This picture is in no way borne out, as we can all testify from our own experience. On the other hand, the Ardrey-Lorenz side cannot be excused from at times almost taking pleasure from painting the blackest possible picture of man—the form of life not only unique in its intelligence and manual dexterity, but equally unique in its moral depravity.

There is agreement that man should be altruistic, but there is disagreement whether or not he basically is. Despite all the whitewashing of the cultural depravity fans, reality seems to show that man's aggression is deep-rooted. Lewis and Towers deplore the fact that so much of the modern literature accepts man's depravity.[8] But to not accept it, is, at present, to simply disregard all the facts of human history.

The question of the age of human aggression is intimately linked to the previous question. Ardrey[9] and Lorenz[10] claim that human aggression is old and dates back millions of years. The imperfect record is believed to demonstrate that Australopithecus killed fellow members, and Peking Man ate his brethren. The opponents argue that the fossil evidence can be interpreted differently and maintain that man the hunter-gatherer was a cooperative, peaceful being—the *noble savage*. They argue that early human aggression would have been detrimental to his existence and led to extinction. In their view, aggression became increasingly prominent after the development of agriculture with the concurrent introduction of property. For them, the human territorial imperative would be a newly acquired trait, rather than an old animal inheritance.

For Ardrey, even man's early tools are in most cases weapons rather than tools. We are all familiar with the fact that many seemingly innocent tools do, in fact, make good weapons—any homicide squad can attest to this fact. Resolution through the fossil record would do much to clarify matters. If, indeed, early man killed his fellow man, then the concept of the noble savage would lose much of its attraction.

Reading the literature, one cannot help but feel that the debate has a great emotional content. To endow man with what has been called a *killer instinct* is degrading, it moves him closer to the animal world—which, of course, is not true, since animals are not generally aggressive—and it accepts his obvious depravity as a God-given fact. It seems that this argument is fallacious and still suffers from the attempt to see man as the centre of the universe and the ultimate in creation. In fact, man's depravity is unique. Presently, he not only kills his fellow members, he enjoys torturing them, which serves no useful purpose at all. Yet human history, both past and present, is full of such accounts. In this regard, it seems that man is unique. To call him a beast is an insult to any self-respecting animal.

In conclusion, one can only say the following. The reading of Ardrey's trilogy[11] makes more sense to the uninitiated (although a natural scientist) than Montagu's *Man and Aggression*.[12] True, Ardrey the playwright does jump to conclusions and often does not stress sufficiently the alternatives. However, to say that his bibliography contains few reputable references is simply not true.[13] Checking reveals that references are properly quoted. What more do you want?

Ardrey also takes credit for unearthing Eugène Marais' work, a South African biologist far ahead of his time.[14] Furthermore, Ardrey has had the courage to integrate a great deal of different material—a task most necessary in our times (see Chapter 12). To demand that such a task be perfect at the first attempt is asking the impossible.

In terms of Lorenz' *On Aggression*,[15] again, much good is contained in it. To a priori declare it is not permissible to compare man with other forms of life is both false and arrogant. Whether Lorenz went too far with the interpretation of his observations on birds remains to be seen. But the idea that there is some message in the study of animal behaviour—and not only that of primates—seems only natural to those who acknowledge the unity of terrestrial life.

The notion the social scientists have of man's infinite malleability is directly contradicted by the last 10,000 years of history, which show no evidence to that effect. Man's foremost thinkers have long recognized and deplored his depravity. Our playwrights have held the mirror in front of our noses for 2,000 to 3,000 years, and the all-learned "animal" has learned nothing. One may, of course, argue that there simply has not been enough time. But if the learning process is that slow, if the cultural evolution is unable to produce what on the human timescale would be called a revolution, then one can indeed condense the future of mankind into "*Homo sapiens,* the stupid s.o.b., has had it."

9

THE MOST DOMINANT SPECIES EVER

We have already alluded repeatedly to our dominance (see Chapters 5 and 6) and it would now seem appropriate to take a closer look. While there is no need to support the above statement, we may look at why this is so and what possible consequences may result from it.

Our dominance results from the fact that we are so adaptable. We live anywhere on the land, in the air (even in space), on the sea, and lately even in the sea. No other form of life is so widespread. In turn, this means that we affect every form of life. If you happen to be an animal or a plant there simply is "no place to run." We tolerate no useless or harmful (to humans) forms of life. Our superiority is complete, and our recognition of this fact permeates our personalities, clouds our thinking, and confirms our idea that the world was created for the express purpose of our enjoyment—that we are indeed above it all. In fact, we are more like visitors, irresponsible ones at that—from outer space—who exploit this place while it lasts, and then move on not caring what, if anything, we leave behind.

Our complete dominance is closely related to our success as super toolmakers. This is responsible for our rapidly increasing numbers and our ability to move into every corner of the world, relentlessly pursuing all that creeps, runs, flies, or swims. More recently, we have come to realize that our conquering all environments is not without costs. It takes an enormous amount of energy to settle man permanently and under modern conditions in any but a moderate climate. Since the presently known major sources of energy are finite, this in itself is a sign that our present dominance can only be a temporary affair.

The consequences of our dominance, as we see it today, are not pleasant to contemplate. Clearly, our presence has already created what later geologists will view as a break in evolution (see Chapter 4). To create such an instability is risky, since one can easily be caught up in it oneself. Given the fact that our presence tends to destroy the diversity of life, and with it life's general chance for further development, Mother Nature must consider us a failure and condemn us in the interest of the persistence of life. To tolerate the worldwide dominance of man for too long could, in fact, seriously threaten the process of life as such. Local dominance, as it must have occurred in the past, never carried that threat with it. It resulted in local extinctions and changes, but never did it affect all environments.

Thus, to press our dominance to the hilt must be viewed as violating an important law of evolution, which Simpson formulated as follows: "One thing that is definitely known now is that breeding for uniformity of type and for elimination of variability in the human species would be ethically, socially, and genetically bad and would not promote desirable evolution."[1] What is true for one species is also correct for life in general. The elimination of variability as we currently practice it is a threat not to one single species, but to terrestrial life as a whole.

TODAY

10

MODERN MAN'S PREDICAMENT

M odern man's predicament—a term often used but seldom well defined. Since we cannot stay away from
it altogether, it may be useful to clarify its meaning.
Commonly, this refers to man's collision course with nature—
his attitude of ruthless exploitation leading to the depletion
of non-renewable resources, the pollution of the environment,
overpopulation, with its accompanying increase in human
aggression, and other effects. This predicament is a result
of the options provided to man by his technology. It is the
technological-scientific revolution that has brought about
the predicament and made the current situation unique in
human history.

Human predicaments have always existed and in themselves are not new. Failure to solve such predicaments always
had disastrous consequences, as history tells us. Cultures
have risen to dominance and disappeared, often without a
trace. Modern man's predicament is different, in that failure
to solve it will result in a mass disappearance of unheard of
proportions. It further entails the decided risk that mankind
as such may be wiped out, even though this possibility may
be overestimated by many.

Also, modern man is more conscious of his place in nature and the universe. Recent research has made him recognize the process of evolution and his position on the earth. He should be more aware of the possibilities of a species such as *Homo sapiens*. Thus, one may expect a greater concern for a problem that could be the final chapter for a species that has only evolved in the last few million years. Earth history tells us that this is not a proud record of persistence.

11

THE TWO CULTURES

The intent here is to discuss the two cultures of C.P. Snow.[1] I am neither qualified nor interested in passing judgment on Snow's general qualifications as a writer. It is not Snow as such that is under discussion, and it is about time we recognize that numerous people have made only one major contribution in their lives. The rest of their works may only represent mediocrity and not be indicative of the one great success. Whether or not this is the case for Snow, I am not to decide, and whether such a performance entitles the creator to eternal fame is also a moot question in the present context. The significance of C.P. Snow (Leavis[2]) is therefore immaterial to our considerations.*

We shall proceed by "reading" Snow's address on the two cultures, briefly contemplate some of the profound statements in this opus, and pick some priorities for more extensive discussion and reflection. The choice, of course, always represents the personal bias of the reviewer.

---

* See Note 32, at xxxiii: "Leavis's contempt was total...Snow 'is intellectually as undistinguished as it is possible to be'."

59

A Quick Glance at the Two Cultures

The paper takes its title from the first of the following four parts:

> The Two Cultures
> Intellectuals as Natural Luddites
> The Scientific Revolution
> The Rich and the Poor

C.P. Snow delivered this address as the Rede Lecture in 1959.[3] It has lost none of its relevance,* but in fact has moved more into focus since that time.

The Two Cultures

In this section, Snow points his finger at, and deplores, the division of human endeavour into the broad fields of the humanities and the sciences.† What Snow does not mention, possibly because it is too explosive a topic, is the fact that scientists and their practical arm, the engineers, have been fantastically successful, whereas social scientists can only be classified as failures in light of human behaviour over the last few thousand years. Simpson has phrased it as follows: "The inequity of knowledge is in itself unethical and is one of man's great blunders. It could be his last."[4] We shall return to this topic (Priority No. 1).

Snow says, "*Each of us is alone*: sometimes we escape from our solitariness, through love or affection or perhaps creative moments, but those triumphs of life are pools of light we make for ourselves while the edge of the road is black: *each of us dies alone.*"[5] (emphasis added) The latter statement caught the wrath of Leavis.[6] The first statement, evidently far more dev-

---

* Humanists seem to dislike this term. According to the dictionary, "relevant" stands for pertinent, applicable, etc. In other words, a relevant topic calls for action, be that consent, continuation, implementation, or dissent and rebuttal. Acceptance of relevance cannot result in indifference, one of the curses of our time. I shall continue to use the term.

† Personally, I classify the engineers and social scientists as the applied scientists and humanists, respectively.

astating, received little attention. One can only wonder why dying alone should be worse than living alone—must have something to do with the humanist's paranoia with death (see Priority No. 4). Of course, this view destroys all that is holy to humanists and social scientists—the view that man is noble, considerate, passionate, and virtuous. We shall not return to this remark, but it deserves scrutiny from all of us during a quiet moment.

Much of the first part of Snow's lecture is naturally devoted directly to the topic of the title—the fact that the species of *Homo sapiens* is divided into two subspecies, that of *man the toolmaker* and *man the contemplator.* Snow seems to consider the narrow-mindedness of the scientists as well-established and thus concentrates on demonstrating that humanists suffer from equal deficiencies. The latter consider their lack of scientific knowledge rather inconsequential, since in their view spiritual matters rank far above technical ones.

In a world where the future of mankind depends as much on technology, and in particular the type of use we make of it, as it depends on spiritual and moral values, such an attitude is indefensible. Naturally, for Snow to pronounce the humanists as equally guilty did not sit well with them, and major attacks on Snow were inevitable. Snow in no way intended to degrade the humanists, he only claimed that the cry of *mea culpa* must ring equally loudly from both camps. We shall leave the topic for the moment and return to it at a later stage (Priority No. 1).

Intellectuals as Natural Luddites

Snow makes the following, very profound, statement: "In fact, those two revolutions, the agricultural and the industrial-scientific, are the only qualitative changes in social living that men have ever known."[7] This we *must* come back to later (Priority No. 2).

The Scientific Revolution

Snow divides what is commonly referred to as the industrial revolution into two parts: the industrial revolution from about

1750 to just past 1900 and the scientific revolution starting about 1920.[8] We shall include the industrial revolution in our discussion of revolutions in general. It seems, however, that this further subdivision is both unnecessary and confusing. We shall not adopt it.

Snow provides a brief but relevant discussion of the fate of the Republic of Venice.[9] Since it amounts to doomsday prophecy for the affluent western societies, it has been largely ignored. After all, we have never learned anything from history,[10] and we still don't for that matter. We shall return to this idea later on (Priority No. 2).

The Rich and the Poor

Snow states: "Life for the overwhelming majority of mankind has always been nasty, brutish and short. It is so in the poor countries still."[11] Naturally, one gets static for such a blunt confrontation with reality. One just does not remind the affluent western world of such unpleasantness. To invade the western ivory towers with such facts of life must be termed almost criminal. How are our intellectual giants to research the fate of humanity, when their peace of mind is shattered in such rude ways? Obviously, Snow must be a man without manners. We shall leave this one for your personal contemplation and turn to another problem.

This whole section of Snow's talk is permeated by the view that the gap between the rich and the poor cannot be maintained. Most of us would agree that in the days of mass media and worldwide satellite communication, this is a reasonable concept. What, however, is open to question, and in my view is outright erroneous, is the idea that we shall all be rich thanks to the industrial revolution spreading to the Third World. This misconception we shall definitely want to challenge (Priority No. 3).

Further on, we find the eternal question: "Is man good or evil?" Snow phrases it this way:

> On the one hand, it is a mistake, and it is a
> mistake, of course, which anyone who is called

realistic is specially liable to fall into, to think that when we have said something about the egotisms, the weaknesses, the vanities, the power-seekings of men, that we have said everything. Yes, they are like that. They are the *bricks* with which we have got to build, and one can judge them through the extent of one's own selfishness. But they are sometimes capable of more, and any 'realism' which doesn't admit of that isn't serious.[12] (emphasis added)

Question: "Is Snow right? Or does he just lose his nerve at the end, not wishing to confront his audience with one more intolerable thought?" Obviously, taking off from the above statement it would not be difficult to write a thick volume (Priority No. 4).

Priority No. 1: The Two Cultures
It has been said that the term *culture* is a poor choice on Snow's part for what is to be expressed.[13] There can be no doubt about its meaning in the context as used by Snow. The topic is too important to get sidetracked into a debate on semantics. We shall not tamper with Snow's original choice.

Question No. 1: Is Snow's claim true, that we are a "split" society? No one in the western world can possibly deny him that he is right. The Americans have perfected the training of the specialist, the *Fachidiot* of the Germans. The rest of the western world, enthralled by American leadership, was eager to follow suit. The result—we are burdened with a class of academics so narrow-minded that they are not only useless, but outright detrimental to the further continuation of western culture, if not humanity as a whole.

Oh yes, we claim to recognize our mistake. INTERDISCIPLINARY STUDIES is written in capital letters not only at the University of Calgary. But what do such studies really entail? The rushing in, one after another, of the various specialists giving their opinions. There is no debate—no listening to one another. After all, what's the point, since who but for example another economist is to challenge the concept of inflation as basically invalid and morally corruptive? Geologists must concern themselves exclusively with rocks—anything else is beyond their grasp. Unless you have been certified by a PhD degree, your view in a field of human endeavour cannot possibly count for more than the uninformed blabberings of an amateur. Robert Ardrey is, at least in part, a typical victim of this thinking. It is totally unacceptable to me.

Question No. 2: Is the formation of the two subspecies of human beings a new trait? Is it a direct consequence of the industrial revolution? Snow did not raise this particular question, but it has been posed by others. I am not at this time in a position to give an answer, but what a beautiful and significant research topic for a historian. Far more relevant than exploring the frustrations of Napoleon between the ages of 6 and 12.

Question No. 3: Is the existence of these two cultures as disastrous as Snow wants us to believe? No doubt about that. In fact, since 1959, the handwriting on the wall has grown both in clarity and in size. Only the most feeble-minded can still overlook it. In the last two decades, no book of significance has been published on either side of the Iron Curtain that can predict anything but doom for mankind on its present course.[14] A little thought will quickly convince anyone that the fundamental wrongs of man are directly attributable to his specialization.

Simpson says, "In the last analysis, personal responsibility is non-delegable."[15] Specialization is on a direct collision course with this profoundly true statement. Scientists are not responsible for the social consequences of their inventions (science is neutral!). Humanists are equally irresponsible, since the basically dirty, applied science is best ignored. The less one knows about it, the better. Maybe it will go away (which would, however, be most deplorable since the material amenities are so enjoyable). The personal responsibility of citizens and parents is delegated to the police, teachers, social workers, or in general simply to that anonymous entity known as *The Government.*

Snow states: "Closing the gap between our cultures is a necessity in the most abstract intellectual sense, as well as the most practical. When these two senses have grown apart, then no society is going to be able to think with wisdom."[16] He also states: "We have very little time."[17] We cannot disagree with these statements and can only deplore the narrow-mindedness of the academic community, which now for two decades has managed to ignore these most urgent pleas, made not only by Snow but by others as well.[18] See Chapter 13 for a further discussion of the gap between the two cultures and its significance.

There can be no question that the human mind is a marvel, but at the same time it is not quite as grand as we make ourselves believe. Thus, we always need *crutches* in order to understand certain concepts.

Paul Schau[19] has provided a beautiful crutch for explaining the multipart truth. Let us assume that truth is given by the following equation:

$$(\sqrt{2})^2 + (\sqrt{3})^2 = 5.000...$$

Let us also assume that we know: $\sqrt{2} = 1.414$ and $\sqrt{3} = 1.732$. This results in 4.999 and the "truth" is quite closely approximated. Let us now assume that we concentrate all our effort on $\sqrt{2}$, such that the square of that number is indeed 2.0000... At the same time, however, we may pay the price for our folly. Since our resources are always finite, we have to neglect $\sqrt{3}$, such that it is only known to the first decimal—yielding a square of 3 equal to 2.89 (i.e. 1.7×1.7). The result is:

$$2.000... + 2.89 = 4.89$$

and we are now much further from the "truth." A simple way to demonstrate that a multipart truth can only be closely approximated (it can never be reached) by equal treatment of all its components.

Priority No. 2: Reflections on Revolutions

Snow's statement that man has only experienced two revolutions that have truly affected his lifestyle is a profound one and in contradiction to what most textbooks have to say on the subject. The term "revolution" evokes in all of us visions of rubble, destruction, decaying bodies, and the accompanying stench. Our thinking is directed to barricades, flags, gunshots, and the final establishment of fairness and justice. The reality, of course, is quite different. In a nutshell: "One son of a bitch takes over from another." And the misery continues. Distress and violence reach a peak during the "revolution," then gradually subside until complacency and boredom take over, and *Homo sapiens* is again ready for some more action. See Chapter 46 for a further discussion of revolutions.

This then has led to the earliest conclusion that history repeats itself. Nothing is further from the truth, as any geologist or, more properly, paleontologist can affirm. Evolution seems to have no particular aim and proceeds on an erratic course. Nonetheless, it does not repeat itself and zigzags to both greater variety and higher complexity. The previous erroneous conclusion is reached simply because an insufficient yardstick is used. Man's recorded history may seem long to him, but it is really no more than a fleeting few minutes (see Figure 3.1) and does not represent a proper sample for such basic conclusions.

No, political revolutions have never accomplished anything, even though this may come as a surprise to many social scientists. Snow is quite correct, the agricultural revolution some 10,000 years ago and the industrial revolution during the past 200 years are the two events that have profoundly shaped man's position on earth (a third candidate may be the taming of fire - see Chapter 46). Of course, closer inspection will show that both revolutions are part and parcel of one development, and the industrial revolution is no more than the steep part of an exponential growth curve (see Figure 14.1).

Some 10,000 years ago (the number tends to move backwards as our knowledge increases), animal power was harnessed, fields were cultivated, and cities began to form. Then, long before Christ's birth, the wheel was developed, and boats, in particular sailboats, became the mode of transportation for both people and goods. Gunpowder was invented and immediately put to use to wipe out fellow competitors.

No, the happenings of the last 200 years are nothing more than the logical consequences of what started long ago, ever since the first stone tools were flaked, and that is more than three million years back, or about 120,000 generations in the past. How do you enjoy digesting that? Bon appétit.

As soon as man emerged as he is today—the product of his mind and his hand—the die was cast. Do we learn nothing from history? Of course not, how could we? Our historians are preoccupied with the classic period some 2,000 to 4,000 years ago. By that time, the course was set and man proceeded along a path 180 degrees off true course, i.e. the course that would serve his long-term interests best.

To learn more about man, and where and when things have gone wrong, one must dig much deeper into the past. Into paleoanthropology, the true emergence of man, and his becoming the number one animal—which from then on has led to his inevitable and total dominance (if you are not a *Homo sapiens*, there is no place to run, swim, or fly). Teilhard de Chardin has said that evolution represents the rise of consciousness.[20] Ergo, man is the ultimate consciousness on earth. Consciousness means development of the ego, and we have come to realize that this is not an unqualified blessing.

To believe that a political revolution has ever brought about a real *qualitative* change in man's life is naive. None of these upheavals has ever changed mankind's direction, only individual fates were affected. The world's major religions have been unable to cope with human nature. Some individuals, rare ones, were gifted enough to see the wrongs and raise their warning voices. But never were they listened to, and never as a species have we been able to "break the pattern," as Snow calls it.[21]

It is clear that salvation from major disaster (left unspecified) can only come from such a breaking of the pattern—a human revolution (see Chapter 47). Is such a drastic change in human nature possible? Studies of the last few thousand years suggest it is unlikely. Such a revolution does not depend on a new system, it depends on a new element—a new individual (the "brick" of Snow). The fundamental mistake of the social scientists is their preoccupation with systems. All human systems—political, economic, and religious—have the capacity to function. The fact that none of them does function is not the fault of the systems but rather due to the fact that they all must rely on the same deficient "brick"—*Homo* not so *sapiens*.

Priority No. 3: Can We All Be Rich?

Snow is not the only one who takes the view that the gap between the rich and the poor is untenable. It is so from a moral point of view but more practically (morals always have had and certainly have today a low market value), it simply cannot be maintained in a world where the mass media let everyone look right into the Shah's dining room. The have-nots are not going to accept their lot as inevitably as they have in the past.

If the present disparity cannot be maintained, then there are basically three alternatives: (1) we are all going to be rich; (2) we are all going to be poor; or (3) we are going to meet somewhere in the middle. The rich are in agreement—at least officially—that the gap must be closed. They envision this closure taking place by bringing the Third World "up to par," thereby creating new customers and "keeping the business going." One does not have to be a mental giant to figure out that this is not what will happen in light of the fact that the total amount of "goodies" (oil, coal, ore, potable water, arable land, etc.) on this earth is limited. In this respect, see the concept of *effective population* discussed in Chapter 15.

The panic of the West with any erosion in the *standard of living* stems from the erroneous assumption that the *quality of life* (happiness?) is linearly related to the standard of living and will grow ad infinitum as the material situation "improves." A more realistic view is shown in Figure 27.1. Unlimited material wealth is no longer a blessing, and one cannot help but feel that in the West we are "over the hill." Western populations as a whole do not, however, realize the folly of their material expectations, and any decline in the standard of living is likely to lead to total chaos, with the collapse of all social order as we know it today. It may be a consolation that, as Simpson once said, one cannot predict the future but only explore its possibilities.[22] The above outcome is no more than one possibility, but, I fear, a serious one.

Priority No. 4: *Homo Sapiens* – Good or Evil? A Little of Both? Much of One and Little of the Other?
To read biographies has its fascination. What one normally discovers is that all the giants of the past, with their wonderful ideas about mankind, were pretty human when it came to their roles as parents, citizens, or colleagues. The present writer is no exception. The cause: *Reason always stops at one's own doorstep.* Lorenz has said that the first commandment should be: "Thou shalt not cheat thyself."[23] Well said, hard to disagree with, but unfortunately even more difficult to put into practice. So far, success has been lacking. The present state of "education" will do nothing to rectify the deficiency. For this, the academics, those exposed to an overdose of modern education, are the living proof. They excel in self-delusion, more even than the so-called man-in-the-street, whoever he is.

Ardrey describes the function of man in terms of: *identity, stimulation*, and *security* (in that order).[24] The first, and most important, leads directly to the concept of the *central position*—the fact that man has seen, and continues to see, himself in the centre of time and space, as a species, as a group, and as an individual. To find that our position in regards to our solar system, our galaxy, and the universe as a whole is *way* off centre was and *is* a blow to the ego that we still have not overcome (see Chapter 2). Equally, to recognize that earth history is possible without the blessed presence of man (see Figure 3.1) is not something we like to be reminded of—a distinctly unpopular thought.

Our more profound thinkers have been well aware of this problem for some time. To quote one more recent proponent

(Toynbee), who phrased it very well: "Every soul, tribe, and sect believes itself to be a chosen vessel."[25] On the other hand, it is interesting to note that the very same author, blessed with such deep insight, got his own priorities mixed up when writing his last book entitled *Mankind and Mother Earth.*[26] Clearly, the title should read the other way around!

The concept of the central position is deeply ingrained in the individual, group, and species. It is the price of consciousness, which is obviously not all good. The concept is also on a direct collision course with the population explosion, creating a loss of identity which is responsible for many acts of foolishness by modern man.

I have stated, and stand by my guns, that the best translation of *Homo sapiens* is "son of a bitch"—rude, not very flattering, and quite obviously totally unacceptable since it is lacking in scientific rigour. Examine your first reaction: Is it? Human history is an unbroken chain of misery, tears, rubble, and violence—practically all self-inflicted. Contrary to common belief, Mother Earth has been kind to us. Sure, occasionally we get decimated by floods, landslides, and earthquakes. However, that toll is infinitesimal compared to the one we inflict upon ourselves (another research topic for historians).

Such a gentle soul as Toynbee, in *Mankind and Mother Earth*—where all human history rolls by in 600 pages—cannot chronicle such history without describing on almost every third page a major population centre getting "sacked," not by tsunamis, hurricanes, or earthquakes, but by human competitors. I invite you instant wizards of the generation under 40, of the affluent technocratic world, to contemplate for a moment Paris, Chicago, or Ottawa getting sacked.* I submit to you: "You ain't seen nothing yet!"

There can be no doubt that our nature has been our worst

---

* For the West, a sacking in Saigon, Lhasa, Beijing, or Irkutsk is quite acceptable. This would satisfy our need for stimulation and could be viewed via satellite on television. It is still a cherished, but unfortunately erroneous, view of the West that all future bad things will conveniently happen "elsewhere."

enemy over the last few thousand years. As long as the "bio-degradable nitwit" (term courtesy J. Hopkins*) called *Homo sapiens* was technically underdeveloped, he was one with all life and, in common with all other forms, could only cause local damage. That is no longer true today. The next round of "sacking" may well be global.

The outsider is always puzzled by the preoccupation of humanists with death. Death is the logical consequence of birth. It is as unpredictable as it is inevitable. Why all the fuss about it? Quite clearly it, more than anything else, threatens our central position. The thought of only being a *tourist* on this earth, to face up to the fact that the personal position—hard fought for—in the pecking order is only of temporary significance, is absolutely shattering to the over-inflated ego.

A more realistic acceptance of the temporary nature of man's stay on earth might go a long way to eliminating one of his major character deficiencies—greed. Greed just makes no sense when all possessions have to be surrendered in what is, after all, a short time (the geologist speaking). Death, of course, we again have in common with all other forms of life. It "lowers" us to the animal level, and that raises another question.

To demonstrate the ignorance of the humanists, Snow chose his examples mainly from the field of physics. I think he missed the boat. There is a field far more important, that philosophers and humanists in general have completely over-looked. It is our relationship with the rest of the animal world. Genetic research has overshadowed all other facets of recent biological research, and that is deplorable. The new studies on animal behaviourism are not only fascinating, but extremely relevant to anyone who seriously wishes to understand the phenomenon of man. We owe Robert Ardrey a great debt. Through his popular writings he has moved animal behaviourism and paleoanthropology into the limelight.

---

* A more refined soul (the reference escapes me) called man a halfway house between the great apes and *Homo sapiens*. Anonymous was charitable.

GREED - THE FACELESS GODS OF MAN

There can be no doubt that man, who is a mammal, is part of the terrestrial animal world. His brain *and* his hand set him apart from all other forms of life, but he is not entirely divorced from it, and he is not unique in an absolute sense. He needs food, shelter, a breathable atmosphere, and sleep—all in common with all other higher forms of terrestrial life. He is superior and more adaptable because of his brain capacity and the skill of his hand. The latter has been referred to as the most marvellous biological tool.

Man is *the* toolmaker—his triumph and his downfall (how come we despise manual work as much as we do?). It has been pointed out that in terms of brain capacity, the mammals that took to the sea, such as the dolphins and whales, may well be equal to man.[27] However, as we noted in Chapter 5, what kind of competitors can these poor creatures ever be, equipped with only flippers?

Writers and humanists discussing man cannot avoid mentioning some of his less lovable traits. These flaws are usually downplayed and referred to as "the beast in man." Anybody writing in this vein in our times has put the stamp of an ignoramus on himself. Man tortures his own kind customarily and en masse, even today. He also kills his own kind in large numbers and gets more efficient at that task all the time. Lorenz's statement that animals do not do this[28] still stands, although accidents can happen. Modern observations on the whole confirm that animals of the same kind do not usually fight each other to death. Earlier observations in zoos (artificial environment) showed a different picture, but are naturally irrelevant.

To burden the rest of the animal world with all that is imperfect in man is not only unfair, but dangerously erroneous, in that it stamps man's depravity as an aberration rather than a fundamental quality. Painful as it may be, it now seems that at least the scientific world is ready to accept man's depravity as a fact. However, as we noted in Chapters 7 and 8, the battle rages as to whether this is the result of our biological evolution, i.e. an inherited characteristic, or whether we have gone astray as a result of the much more recent cultural evolution. Proponents of the latter theory are those who believe that man is infinitely malleable, entirely a product of his environment (the cultural evolution), with biological traits, or instincts, playing but a negligible role in his behaviour. As a natural scientist, I find the more moderate view of Dobzhansky, who sees man as a product of "nature and nurture" (the percentages left open but neither of them negligible),[29] more acceptable.

It is clear that only the cultural evolution proceeds at the speed necessary to save mankind. The solution—the human revolution—can only be produced by rapid cultural evolution (see Chapter 47).

As we noted in Chapter 8, if our deficiencies are largely biological, as Ardrey and Lorenz would have it, then the cultural human revolution must override inherited biological characteristics. If our depravity is a result of our cultural evolution, and the *noble savage* of Rousseau is a fact,[30] then we must reverse this evolutionary trend. In either case, the task may prove too much.

The original question (which by now we have forgotten) was: Is man good or evil? Let us first define *good* as beneficial to the further existence of mankind and *evil* as detrimental to that goal. As realists, we must admit that by that definition, man is more evil than good. To deny him any good is to bury hope—to see him as all good is to commit suicide (see Chapter 38 for a further discussion of the concepts of good and evil).

Conclusions

1. The two cultures are a fact of modern life. The excuse that the world is too complex for one individual to understand is unacceptable and no more than an expression of man's inherent intellectual laziness. As long as this state prevails, no solutions to modern man's predicament are possible. The inequity of knowledge will lead to the downfall of western civilization and in the worse case may, indeed, be a contributing factor to man's premature extinction.

2. Political revolutions are a myth. They have affected individual destinies, but never that of mankind. Social scientists should stop overrating them. All systems have the potential to function, but no system will function as long as the building brick, *Homo sapiens*, remains an unsuitable element.

3. We cannot all be rich. The gap will be closed. The closure will be painful for the rich. In the worst case, it could be fatal for all.

4. The concepts of identity, stimulation, and security are useful in understanding how man functions. The craving for identity (over-inflated collective and individual ego) as well as the need for stimulation (violence, war, etc.)—both are the real threat to man's existence. Their moderation to a tolerable level and/ or channelling into a harmless direction is a must. The task of this *human revolution* is a challenge that can only be met through a combined effort on the part of both the humanists and the scientists.

A more realistic attitude on our part towards the universe, the earth, and other forms of terrestrial life will go a long way towards solving our problems. In our hearts, we still believe that all our surroundings were created for our exclusive benefit. The light does not shine that much brighter in the 20th century than in those past.* Let us not be deceived just because we *have* set foot on the moon. Maybe we never should have reached for the moon and simply left it in poetry.

Post Scriptum
To solve our problems, we shall form the black hole religion. The ultimate fate of the earth: a black hole. The final equalizer—your basic components only millimetres away from those of Stalin, Hitler, Idi Amin, Albert Schweitzer, Leonardo da Vinci, Einstein, Gutenberg, and Ali the Great.

We shall adopt the single commandment: *Thou shalt give your fellow man a fair shake.* We then proceed to immediately pervert our ideas and all will be back to normal.

---

* May I call to your attention that in the modern skyscrapers of downtown Calgary (all post 1960) we may find high-speed digital computers on the 6th floor, the 12th floor, the 21st floor, but never on the 13th floor. Simple, there are no 13th floors. I have questioned several engineers about the structural integrity of these towers. The answer is usually a dirty look. Contrary to popular belief, the superstitions of our forefathers have not been banned to the heart of Africa or the core of the Himalayas.

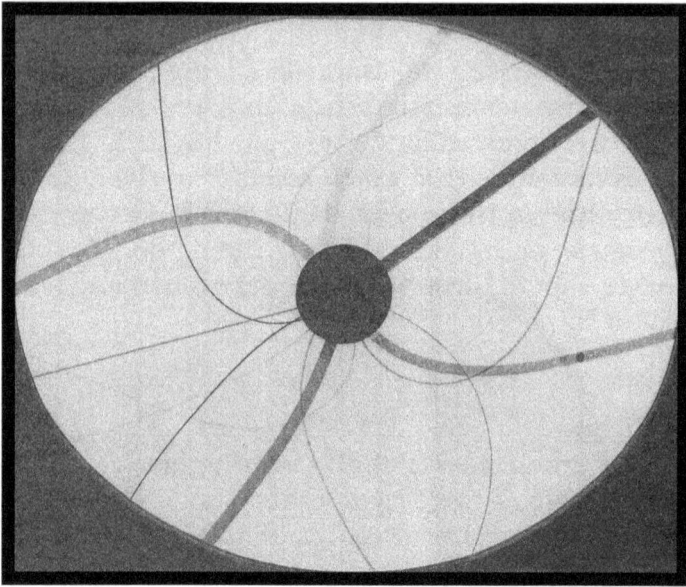

THE EARTH AS A BLACK HOLE (TO SCALE[31])

Editor's Note

"At a few minutes past five o'clock in the afternoon of 7 May 1959, a bulky, shambling figure approached the lectern at the western end of the Senate House in Cambridge."[32] By the time he sat down a little over an hour later, C.P. Snow had "...started a controversy which was to be remarkable for its scope, its duration, and, at least at times, its intensity."[33]

Over 50 years later, we ask ourselves: "Is Snow's thesis still relevant?" John Naughton, Professor of the Public Understanding of Technology at the U.K. Open University, had this to say at a convocation address to the Faculty of Arts graduating class, University College Cork, Ireland, in 2002:

> ...the gap between the two cultures today is as wide as it was when Snow drew attention to it two generations ago.
>
> In fact, if anything it's wider and more dangerous. In the old days, those in the human-

ities culture were merely ignorant or indifferent to science. Now they are openly contemptuous of it. Numerous opinion polls show that public distrust of science and technology has reached a pathological level—to the extent that, in Britain at least, many parents are refusing to have their children immunised using the MMR vaccine despite reputable scientific advice that it is safe, with the result that preventable childhood diseases are now on the rise.

On the other side of the equation, scientists and engineers behave in increasingly arrogant ways—especially in areas like genetic engineering or reproductive science. They assert no boundaries should be placed on their inquiries, regardless of the potential environmental, ethical or social consequences of what they are discovering or inventing...

The consequences of all this is that there is not much communication between the two cultures any more, and what little there is tends to be abusive and unreflective. This dialogue of the deaf will be disastrous for our societies, which depend for their health and prosperity on the judicious, ethical and efficient application of scientific and technological knowledge. We must find a way of enabling CP Snow's two cultures to talk to one another.[34]

How to enable the two cultures to talk to one another? For the author's suggestions, see Chapter 29 regarding a proposal for uniting knowledge at the university level and Chapter 44, where the *Interdisciplinary Dialogue*—described as the "Challenge of our Times"—is explored. A challenge we have yet to meet.

12

ON THE FRAGMENTATION OF KNOWLEDGE

The fragmentation of knowledge is presumably one of the most devastating phenomena of our times. Since I have no intention, as is obvious from the table of contents, of adhering to the concept of specialization, we may as well address ourselves to this problem.

Knowledge is divided between the humanities-social sciences, and the natural sciences and their applied arms, the various fields of engineering. The fact that the world is a whole is totally ignored, and many of our contemporary problems can be directly attributed to this. Not only do we have the above grand split of human knowledge—much deplored by C.P. Snow in his two cultures (see Chapter 11)—but the fragmentation reaches much further down. Every field of specialization, such as civil engineering, geology, biology, and all the others, is further divided. It takes anywhere from 10 to 30 faculty members at the university level to cover all the branches of such fields. And not only will those faculty members hardly ever communicate with anybody outside their specialty, but most often they will not even be comfortable, or feel competent, enough to take an interest or argue amongst themselves. This has dire consequences.

One, of course, is that it destroys the myth of the superior intellect of the academics. Academics are, by and large, simply people that have a particular "knack" for certain things. Often that knack may indeed be remarkable, but more often it is really rather ordinary and leads to work that on the whole is intellectually no more demanding than many other jobs which are not endowed with the same status. The excuse is the often heard view that human knowledge has now reached proportions that simply defy the capacity of an individual brain. It would be foolish to deny that there is a certain truth to this statement. However, often it is simply an expression of intellectual laziness or incompetence. Since the actions of the individuals have an effect on the sum, it is simply not permissible to neglect the whole.

To attempt to assess the total picture takes both a far superior and at the same time far more committed intellect. There can be no question that the task of the generalist is much more demanding than that of the specialist. However, current trends are towards specialization. In fact, the generalist is downgraded as a person "who knows a little about a lot." The prevailing attitude is that unless you have a PhD in a given field you should stay out of it. To pronounce judgment on, say, economic matters as a geologist, is considered pure heresy, and one is told to "go back to your rocks."

Since it is naturally true that the generalist takes a larger and considerable risk in being proven wrong, the safe path is to remain the specialist—to look at a minute fraction of the sum, to rehash the same little piece of know-how for a lifetime—and thereby be a greatly admired expert.

When the need for integration is recognized, some experts get together and throw their individual bits of information into a computer (a process one refers to as systems analysis) and let it sort out the mess. The resulting prophecy, which is totally dependent on the information and instructions fed into the machine, then forms the basis for endless further discussions, which usually get lost in details that are both irrelevant and irresolvable.

There is no doubt that today the world is affected by the

actions of the most dominating species ever, both in terms of numbers as well as in terms of power—*Homo sapiens*. What this species does and the effects of its actions are a function of a number of variables, such as its technological capabilities and its values, which in turn are both a function of the biological and cultural nature of man. To totally divorce the study of these aspects from each other, more than any other single factor, is to invite collective suicide.

Many have warned about the fallacy of the division of knowledge,[1] but no action has occurred. The blame for this inactivity must be placed squarely at the feet of the academics. Any person with 20+ years of training must recognize that one cannot argue against the case of the generalist. To do so is both wrong and immoral.

13

THE CENTRAL PROBLEM:
THE IMBALANCE OF GROWTH

As we noted in Chapter 11, our present lives have been deeply affected by two revolutions in our past—first, the agricultural revolution some 10,000 years ago and second, the industrial revolution during the last 200 years. One may also follow C.P. Snow and separate from the industrial revolution, the scientific revolution of the last 50 years.[1]

There can be no doubt that modern man is the most outstanding toolmaker, outclassing every other form of terrestrial life by many miles. We need not dwell on our successes as technologists. They are before our eyes every day. We only fail to recognize the marvels of modern technology because we become accustomed to them even faster than we invent them and therefore take all these modern wonders for granted. Another fact that distracts from their admiration is the fact that many of these inventions fall into the class of weapons. One can hardly argue with Ardrey that man has always been willing to turn tools into weapons, and that the invention and construction of weapons, even to this day, seems to be more fascinating than the production of mere tools.[2] Be that as it may, the success of the modern engineer remains undisputed.

But man is not only a superior toolmaker. Any humanist and social scientist would shudder at such a one-sided definition. He is also the most conscious form of life. He reflects, thinks in abstract terms, and admires beauty. He is also greedy and egocentric beyond expectations. His actions are governed by his desire for identity, stimulation, and security.[3] And in this regard, there is no trace of any revolutionary movement in the last 10,000 years. We talk boldly about education and educational systems and fail to recognize that nobody has been educated in the last few thousand years. All history tells us is that the basic human motivations have remained unchanged. The quest to "know thyself," so ongoing for all of man's history, has yielded virtually no results except, perhaps, for a chosen few that do not affect the destiny of the human race.

Returning to Simpson's claim that the inequity of knowledge is criminal, we can look at Figure 13.1. If one can define a material affluence index (A_m - see Chapter 15) as a measure of man's material wealth, then one can similarly define a spiritual affluence index (A_s) as a measure of man's wisdom. Due to the agricultural and industrial revolutions on the one hand, and our spiritual stagnation on the other, the ratio of these two indices has followed the curve shown in Figure 13.1. A destitute but wise human being is an evolutionary success, but a wealthy and almost all-powerful moron is doomed to extinction. In the final analysis, every facet of what is usually referred to as "modern man's predicament" can be blamed on man's abundance of technology and lack of wisdom. If the term *Homo sapiens* remains the designation for a mechanical genius and a spiritual imbecile, the fate of the species is, indeed, sealed.

Thus, we are facing the paradox that in our capabilities as toolmakers we have progressed beyond all expectations, while what one might term "human wisdom" has remained stagnant. It is this imbalance in growth that is now threatening our further existence. In fact, the most recent technological developments, such as television, have further enhanced this gulf and led to a stagnation of human intellectual creativity.

This, of course, directly ties in with the oft-heard remarks

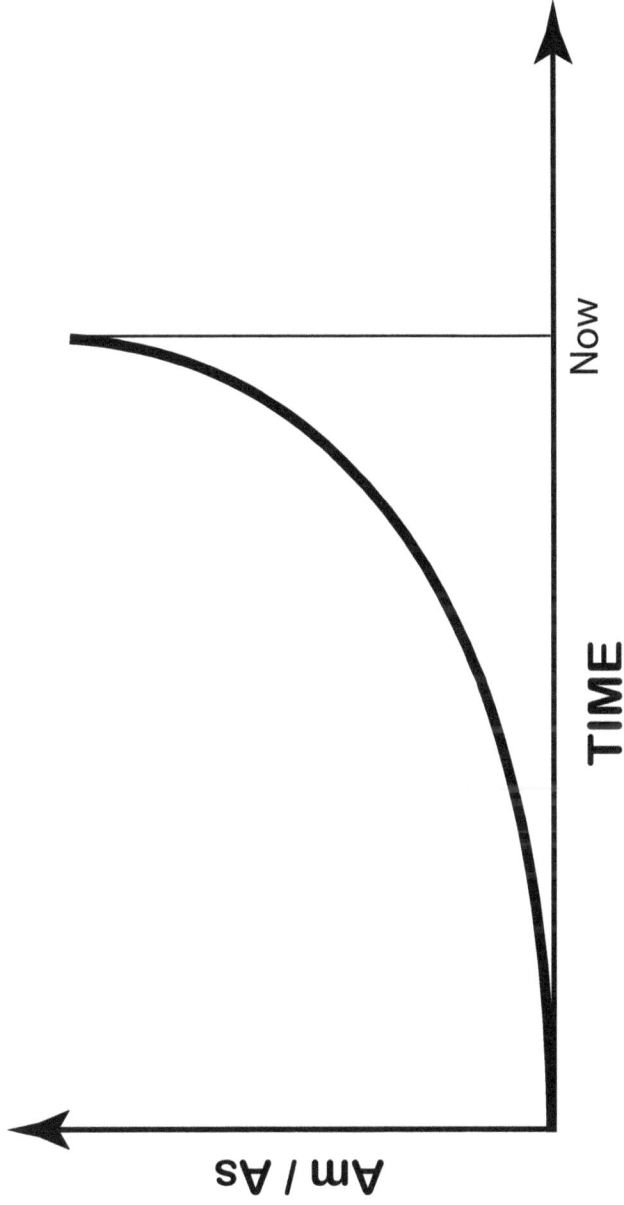

FIGURE 13.1: THE IMBALANCE IN MAN'S DEVELOPMENT. Due to the industrial revolution and spiritual stagnation, the material affluence index (Am) has rapidly increased while the spiritual affluence index (As) has remained stationary. As a result, the ratio of these indices has been knocked out of its equilibrium position.

that all technical developments are neutral, neither good nor bad (guns don't kill people—people kill people). However, it seems that consistently we have managed to make the worst of most of these developments, to the point where what potentially could be a blessing has turned out to be a curse in the long run—and all this because we have permitted the technological skills to be developed ahead of any human skills.

This indirectly indicates that natural scientists and engineers have been an unqualified success as specialists, while the social scientists and humanists must be classified as failures. This does not shift the entire burden of responsibility onto the shoulders of the latter. The engineers have never attempted to make moral decisions about their inventions. They have been fully aware of the use to which so-called neutral inventions have been put. Yet they have been satisfied to yield to the temptation of the technical challenge and to turn the product of their prowess over to the politicians, rejecting any responsibility for the eventual applications.

Thus, we have on the one side, the humanists and social scientists, intellectually and financially underdeveloped, labouring away at the difficult task of understanding and changing human nature, and on the other hand, the engineering expert, the glorified expert of the scientific society, responsible only for technical gadgetry. Clearly, this imbalance in success can no longer be tolerated in the interests of the future of man. Since technical advances cannot be undone, the only viable approach requires a slow-down in scientific advances, with the concurrent commitment of the scientist towards society. The restoration of a healthy balance clearly lies in the hands of scientists, who must slow their own progress while at the same time lend a helping hand in human development in order to achieve this.

14

THE LIMITS TO GROWTH REVIEWED

The *Limits to Growth* by Meadows et al.[1] has caused a great deal of discussion in the past few years.[2] The book has decidedly two parts. The first part introduces the topic of exponential growth. It explains the nature of such growth in simple terms and also demonstrates the impossibility of continued exponential growth in a finite system such as the earth. A typical exponential growth curve is shown in Figure 14.1. Such curves are characterized by the doubling time (t_d), the time it takes to double the initial amount—be that the consumption of energy or other raw materials, the population, or the salary of a tradesman. The doubling period remains constant as long as the growth rate in percent per annum is fixed. It is evident that growth for a long time (0 to t_1 in Figure 14.1) may be almost imperceptible, only to then accelerate (after time t_1 in Figure 14.1) in a most dramatic manner.

The book must be termed trivial in its basic message and subversive in terms of the free enterprise system (and most others for that matter). This does not mean to say that the book is without value. Quite to the contrary, the authors have managed to focus attention on the topic of exponential growth,

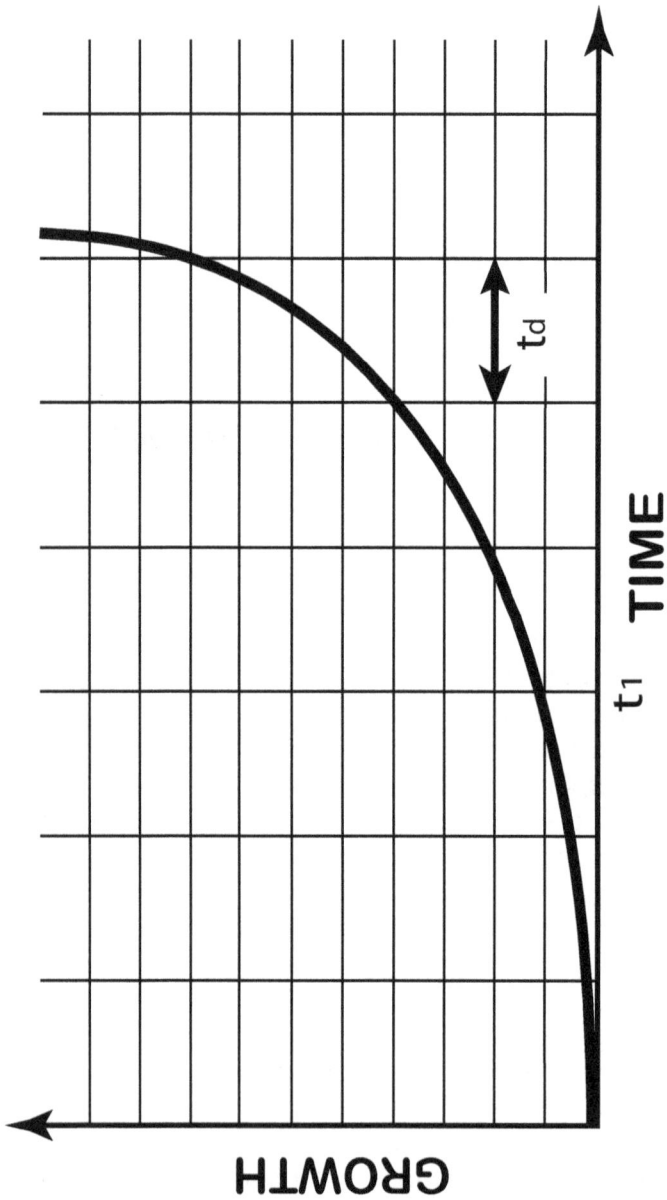

FIGURE 14.1: A TYPICAL EXPONENTIAL CURVE. Note the rapid rise after t_1. The time for doubling the previous amount (t_d) is fixed. 5 doublings result in a factor of 30. 10 doublings yield approximately 1,000. AN INNOCUOUS BEGINNING WITH A DISASTEROUS ENDING!

a most meritorious endeavour. They aptly illustrate the vicious nature of this type of growth with the parable of the water lily pond (there are many such stories). A pond contains a water lily plant which doubles in size every day (doubling time = one day). The pond will be covered in 30 days. At what time will half the pond be covered and how much time is then left to take corrective action, i.e. to cut the lily plants? Answer: On the 29th day and you have exactly one day to save the pond. A most convincing demonstration of the deceiving nature of exponential growth.

The authors provide the following simple rule for approximating the doubling time:

$$t_d = 70/p$$

where p is the percent growth per annum.

Today, growth rates of 2% to 10% must be considered modest in many cases, yet such rates result in doubling times ranging from 35 to 7 years.

Unfortunately, the authors do not carry their basic argument quite far enough. One can easily, without any substantial knowledge of mathematics, show the following:

doublings	1	2	3	4	5	6	7	8	9	10
growth rate	2	4	8	16	32	64	128	256	512	1,024

From which follows the easy rule of thumb:

5 doublings result in a 30 fold increase
10 doublings result in a 1,000 fold increase

We may apply our new-found knowledge to another growth parable in order to gain an even better appreciation for the powerful nature of exponential growth. The chessboard riddle mentioned in *The Limits to Growth* very forcefully brings home both the deceptive nature as well as the powerful expansion of exponential growth. It goes as follows. A Persian courtier presented the king with a beautiful chessboard. The king wished to show his appreciation for the fine gift. The courtier, being a modest yet obviously well-educated man, only asked to be given a grain of rice for the 1st square, two grains for the 2nd square, four for the 3rd, and so on. The king readily agreed to such a seemingly modest request. To make a long story short—the king went broke!

Really? Well, let us just figure out what the king owed our humble courtier for the 64th square. The 1st square is good for one grain of rice, thus by the time we arrive at the 64th square we have 63 doublings. Our rule of thumb tells us that 10 doublings result in a factor of 1,000 or 10^3 and 60 doublings therefore give a factor of 10^{18}. To achieve our 63 doublings, we add a further three doublings with a factor of 8. Thus, to reach 63 doublings, the 64th square is covered with 8 × 10^{18} grains of rice. Not a particularly meaningful number.

Measuring rice, we find that 500 grains corresponds to a volume of 10 cubic centimetres. The 64th square is therefore covered with 160 cubic kilometres[3] or about 40 cubic miles of rice, weighing about 12 × 10^{10} tons. When we consider that global rice production for 1971 was 3 × 10^8 tons or 0.1 cubic miles[4] (i.e. 1/400th of what the king owed), we realize that the king just did not have a chance. Not being acquainted with exponential growth does have its dangers.

Can we disagree with Hubbert who has stated: "It is as true of power plants or automobiles as it is of biological populations that the earth cannot sustain any physical growth for more than a few tens of successive doublings."?[5] One might add that even this view is most optimistic and only applicable to situations where the initial starting amount is exceedingly small.

A 1% per annum growth rate, which in our society would be considered essentially a no growth situation, leads to a doubling in 70 years—one human life span. Nobody can deny that even one doubling creates a basically new condition. A 10% per annum growth rate, on the low side for many items by today's standards, results in a 1,000 fold increase in one human lifetime! So much for the first part of the Meadows et al. publication. It represents a slap in the face of modern western education that it takes several MIT professors and a computer to elucidate such a simple truth and bring it to the forefront. The conclusion is inevitable—present growth rates can be no more than a short-lived, transient phenomenon. This thought is subversive to any economic system that equates *bigger* with *better.* The storm against Meadows and her colleagues is not borne by reason, but by the fact that their ideas are more revolutionary, and a greater threat to the status quo, than any political ideologies.

Figure 14.2 depicts exponential growth in a finite system such as the earth. The realities all look unattractive. Solutions may be found *by* us and result in Curves 2 and 3, or they may be found *for* us (imposed by nature's limitations) and resemble Curve 1. Such a situation, even though the scenario is slightly different, has been described by Christopher in *The Death of Grass*.[6] The Great Stock Market Crash of 1929 also forms an excellent example of such a collapse curve, as does the German inflation of 1921-1923.

In the second part of their book, the MIT authors have applied systems analysis in order to explore the possibilities of the future. They realize that a series of complex relationships exist, such as shown schematically in Figure 14.3. As an example, industrial production leads to both the exhaustion of non-renewable resources and the pollution of the environment. The basic message simply states that tampering with one part of the system leads to repercussions in other parts.

The MIT model considers as individual factors population, industrial production, food production, pollution, and the depletion of natural resources. The detailed world model, which has come under so much fire, does indeed suffer from unrealistic and restrictive assumptions. The most important has been brought into the limelight since 1972, namely the uneven distribution of the world's natural resources—the fact that the *consumers* and the *suppliers* are different people and, to put it mildly, not on friendly terms. Or, brought down to its most simple terms, those who thought they were rich are no

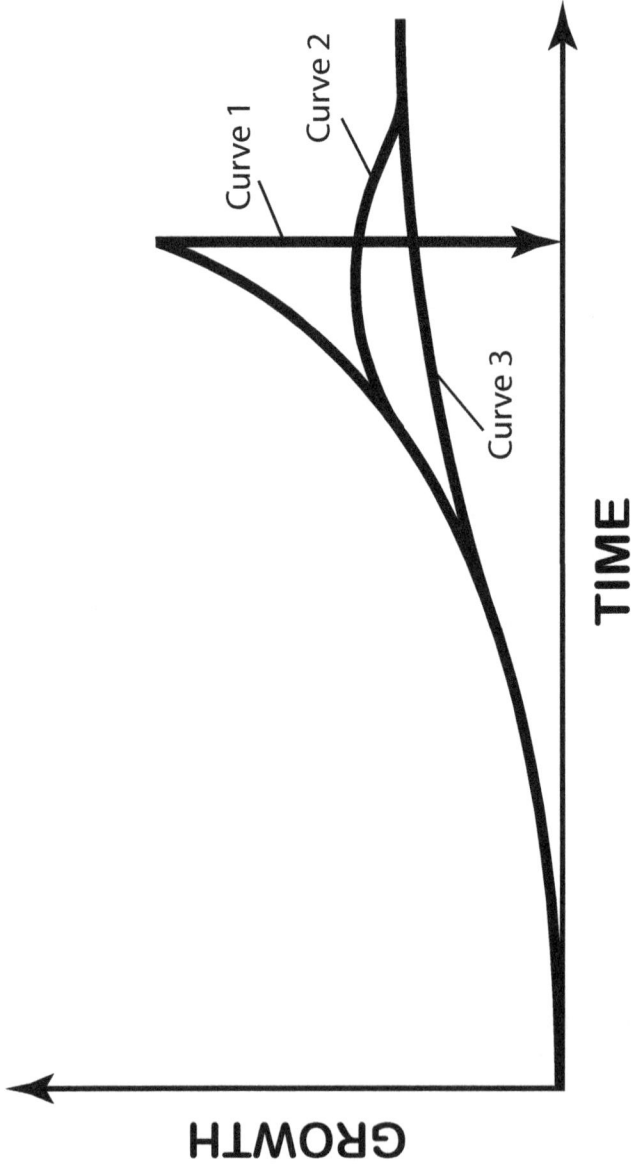

FIGURE 14.2: THE CHOICE – EXPONENTIAL GROWTH IN A FINITE SYSTEM. There are three possible outcomes: Curve 1 – Nature's solution (catastrophic); Curve 2 – Man's solution if he is slow, but acts soon; and Curve 3 – Man's solution if he acts now (best case).

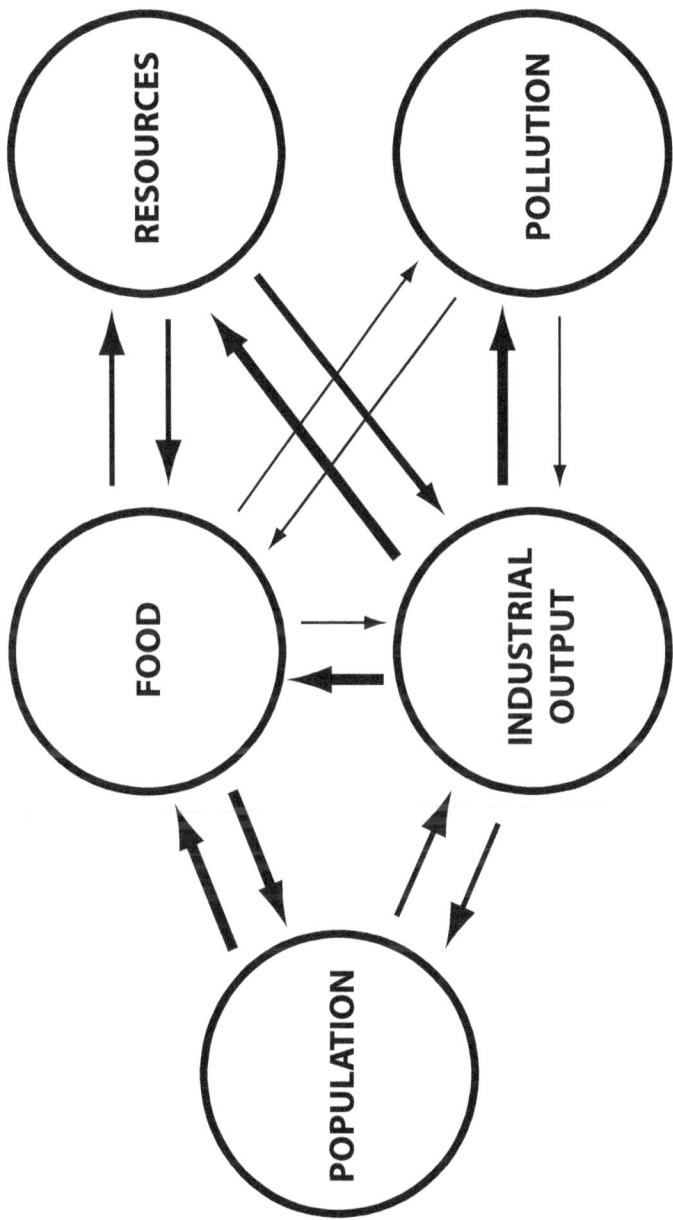

FIGURE 14.3: THE MODERN WORLD - COMPLEX RELATIONSHIPS. The basic message is that tampering with one part of the system leads to repercussions in other parts.

longer so, whereas those that seemed poor are, in reality, in possession of great wealth. However, the authors have been honest in pointing out these shortcomings, and it is therefore improper to accuse them of false prophecy.

The models are supposed to be guidelines, possibilities, rather than firm predictions of the future. It is nonetheless interesting to note that all the detailed models, with widely varied assumptions, predict instability within the next 100 years. This is in full agreement with what a more intuitive assessment of the world situation would suggest. It is the basis for our statement in Chapter 40 that the term *future* ought to be measured in decades.

It should be noted that the MIT authors are, in fact, optimists. Most of their curves take the shape of Curve 2 in Figure 14.2. The idea of a catastrophic collapse, such as shown in Curve 1, and which is a possibility that cannot be ruled out, does not figure prominently in the MIT study. Either it was found to be too shocking or else the computer is incapable of handling discontinuities.

The MIT publication does not spell inevitable doom as suggested by Maddox.[7] All predictions are based on the assumption that no drastic changes occur both in human attitudes and technology. The only optimism that Maddox, or anyone else for that matter, can offer is based on precisely the same assumption—drastic changes in our value systems and basic improvements in technology.

The Limits to Growth has come under heavy fire precisely because it forecasts the doom of the western political, social, and economic systems far more than any foreign ideology. Much of the fire has been directed at the individual models of Meadows and her colleagues. We have already seen that the input assumptions do indeed limit the usefulness of these initial models, a fact well stated by the original authors. Their value as an attempt to predict possibilities for the future, however, remains undisputed.

Certain of the assumptions have a negative influence. But other detrimental effects have not been considered. To say that these models are biased in a pessimistic way is unfair. Without being unreasonable, the assumptions could be changed and produce an even more pessimistic forecast. What is more disturbing, is that Maddox[8] and Beckerman,[9] to name only two, dispute the basic fact that infinity has no place in a finite system. They maintain that growth cannot be stopped, essentially because it is unacceptable to man. The sheer arrogance of this argument is breathtaking. One subtitle in Beckerman's book gives his attitude away: "Man the Measure of all Things." This gets us right back to our introduction and consideration of space and time. To elevate man's desires to a natural law is just asinine. To argue, essentially, that growth must and will persist simply because anything else is unacceptable to us, represents the utmost in arrogance and ignorance.

How people can teach economics at the university level with this type of attitude is beyond imagination. Possibly, the same situation existed in other countries as did in my time in Switzerland. The degree of economics was reserved for those who could not cut it anywhere else yet had to have the status symbol of a doctoral degree. The confusion in economic thinking that exists today leads one to believe that this situation must have been widespread and not confined to tiny Switzerland.

Beckerman, by his attitude, refutes the possibility of what I have termed the *human revolution*. I have no illusions that this is an exceedingly difficult task, as I imply by the name. However, never before in man's history has the pressure to be sensible been as strong as today. It is no longer a matter of seeing the Roman Empire disappear. If our present situation comes to a head, the world population will be reduced by three billion people and possibly four, i.e. eradicated.

One saving grace, which nobody will be able to predict, is where the stage is and where the audience is. The western world still believes that all these unpleasant things will conveniently take place in Africa and Asia and the show can be watched via Telstar in North America and Europe. In the time of intercontinental missiles and satellites, this calculation may well turn out to be a miscalculation.

Editor's Note

The Limits to Growth has continued to attract attention since its publication over 35 years ago. In 1993, the authors published *Beyond the Limits* as a 20-year update on the original material.[10] They concluded that two decades of history mainly supported the conclusions they had advanced 20 years earlier. The 1993 book also offered one major new finding—it was suggested that humanity had already overshot the limits of earth's support capacity, hence the title of the book.

One of the world's leading experts on *peak oil*, Matthew Simmons, looked back on this seminal work in a 2000 Energy White Paper.[11] He concluded that Meadows and her colleagues were essentially right and their critics off-base:

> It would be naive, in my opinion, to assume the gap between rich and poor could stay as it is now, and even more naive to assume this gap can grow without finally creating massive civic turmoil. If the gap gets too great, the poor will finally "come over the walls of prosperity" and attempt to redistribute this wealth. History has shown this to be the case, time after time. Most of our worst wars were not ideological battles but true fights over the redistribution of wealth.
>
> But closing the rich/poor gap needs very careful implementation, as the exponential changes in both energy resources and a staggering number of other factors, including the pollution these increases imply, will strain the world's logistic and resources availability to its limits.
>
> Phase One of the predicament of mankind never really made it to Phase Two. Instead, rather than merely ignoring this work and forgetting its chilling conclusions if the issues raised were forgotten, too many "experts" decided to use this thoughtful work as an easy target of intellectual scorn.
>
> As a serious student of energy for the past 30

years, and a strong believer that compounding historical trends are often a far more reliable way to project the future than any alternate method, the world simply cannot continue the population growth in the poor parts of the world and also have these impoverished people climb the ladder of affluence. The energy usage these numbers imply do not match any sound plan for ever supplying the attendant energy this scenario creates.

Is there time to begin the thoughtful work which The Club of Rome hoped would take place post 1972? I would hope so. But, another 10 years of neglect to these profound issues will probably leave any satisfying solutions too late to make a difference. In hindsight, The Club of Rome turned out to be right. We simply wasted 30 important years by ignoring this work.[12]

The most recent update by the original authors was published in 2004 under the name *Limits to Growth: The 30-Year Update*.[13] Donella Meadows (died in 2001 but contributed to the update), Jørgen Randers, and Dennis Meadows again updated and expanded the original version and offered the following, not entirely pessimistic, prognosis:

We have said many times in this book that the world faces not a preordained future, but a choice. The choice is between different mental models, which lead logically to different scenarios. One mental model says that this world for all practical purposes has no limits. Choosing that mental model will encourage extractive business as usual and take the human economy even farther beyond the limits. The result will be collapse.

Another mental model says that the limits are real and close, and that there is not enough time, and that people cannot be moderate or responsible or compassionate. At least not in time. That model is self-fulfilling. If the world's people choose to believe it, they will be proven right. The result will be collapse.

A third mental model says that the limits are real and close and in some cases below our current levels of throughput. But there is just enough time, with no time to waste. There is just enough energy, enough material, enough money, enough environmental resilience, and enough human virtue to bring about a planned reduction in the ecological footprint of humankind: a sustainability revolution to a much better world for the vast majority.

That third scenario might very well be wrong. But the evidence we have seen, from world data to global computer models, suggests that it could conceivably be made right. There is no way of knowing for sure, other than to try it.[14]

This "sustainabilty revolution" is similar in spirit to the "human revolution" called for in this book (see Chapter 47).

Most recently, in 2008, Graham Turner at the Commonwealth Scientific and Industrial Research Organization (CSIRO) in Australia published a paper titled *A Comparison of "The Limits to Growth" with Thirty Years of Reality*.[15] It also compared actual data from the past thirty years with the predictions made in 1972 and found changes in industrial production, food production, and pollution were all in line with the book's predictions of economic collapse in the 21st century.

The conclusion is sobering—enough to wonder if the question mark in the title of *The Vanishing of a Species?* is overly optimistic:

> Appropriate and publicly available global data covering 1970-2000 has been collected on the five main sub-systems simulated by the Limits

to Growth World3 model: population, food production, industrial production, pollution and consumption of non-renewable resources. In the style of predictive validation, this data has been compared with three key scenarios from the original LtG publication (Meadows at al., 1972). This comparison provides a relatively rare opportunity to evaluate the output of a global model against observed and independent data. Given the high profile of the LtG and the implications of their findings it is surprising that such a comparison has not been made previously. This may be due to the effectiveness of the many false criticisms of attempting to discredit the LtG.

As shown, the observed historical data for 1970-2000 most closely matches the simulated results of the LtG "standard run" scenario for almost all the outputs reported; this scenario results in global collapse before the middle of this century. The comparison is well within uncertainty bounds of nearly all the data in terms of both attitude and trends over time. Given the complexity of numerous feedbacks between sectors incorporated in the LtG World3 model, it is instructive that the historical data compares so favorably with the model output...

In addition to the data-based corroboration presented here, contemporary issues such as peak oil, climate change, and food and water security resonate strongly with the feedback dynamics of "overshoot and collapse" displayed in the LtG "standard run" scenario (and similar scenarios). Unless the LtG is invalidated by other scientific research, the data comparison presented here lends support to the conclusion from the LtG that the global system is on an unsustainable trajectory unless there is substantial and rapid depletion in consumptive behaviour, in combination with technological progress.[16]

15

POPULATION, AFFLUENCE AND THE CONCEPT OF EFFECTIVE POPULATION

Recently, much attention has been focused on the rapid increase in the human population, particularly over the past 100 years. Statistics indicate that on a global scale this increase is still progressing, and the best estimates indicate that the doubling period is in the order of 30 years. The total number of *Homo sapiens* now stands at about four billion. With this number, we are undoubtedly the most dominating species in the history of the earth. However, it is not our sheer number that accounts for our dominance. There must be many species, particularly in the family of insects, which out-number us by several orders of magnitude. What makes our number so significant is the way we live—our technology.

There is no doubt that every one of us, even the most humble, places a burden on the earth. Each one of us makes what one commonly refers to as *demands*. We demand space, food, energy, raw materials and in return place our refuse on the earth. It is quite obvious that these demands vary strongly from place to place and are quite different for a Bihari peasant than for a U.S. suburbanite. Of course, man is not unique in that sense. All living things make demands. However, in the animal world, we can write the following equation:

EQUATION 1: DEMANDS = NEEDS

While the specialists in the field—the biologists—may have discovered some exceptions to this rule, it is highly unlikely that these are anything but true rarities confirming the rule that animals generally only satisfy their needs. Man is unique in the sense that the above equation must be modified for him in the following manner:

EQUATION 2: DEMANDS = NEEDS + EXPECTATIONS

The law of nature requires that the needs of a species are met; otherwise it is condemned to extinction. However, there is no such law that would require the expectations to be met. Expectations are those material requirements which man feels are essential for a dignified and comfortable stay on earth. If some contemporary authors say, "We must meet the needs and expectations of the people," we can only say, "Nonsense." The needs must be met, but the unreasonable expectations of modern, affluent man cannot possibly be met.

In passing, one might note that the exaggerated expectations of *Homo sapiens* are largely a consequence of his greed. It is interesting to record that this seems to be a unique trait. Those who feel compelled to see man as a semi-divine subject might realize that while man is distinctly different from all other forms of terrestrial life, his uniqueness is not in all respects one of superiority.

Figure 15.1 shows the demand curve broken down into its components of needs and expectations. It is quite clear that our main problem is related to the expectations rather than the needs.

Figure 15.2 presents the result of these exaggerated demand curves or what I have labelled: the DISCREPANCY CURVE. Production (supply) of a commodity—be that food, oil, metals, or something else—will at first lag demand. It is subsequently stimulated by demand to where production exceeds demand

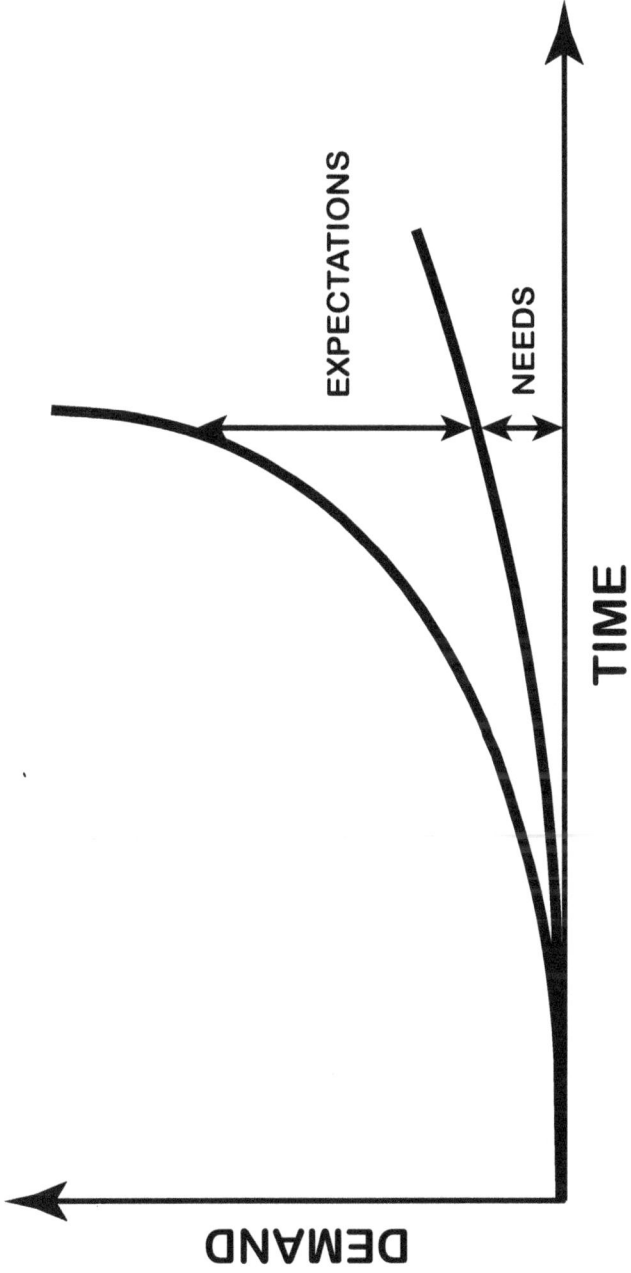

FIGURE 15.1: MANKIND'S DEMAND CURVE - NEEDS AND EXPECTATIONS.
The curve expresses the view that the expectations of the affluent West are unreasonable and still rising. This situation is bound to terminate, the only question:"How and when?"

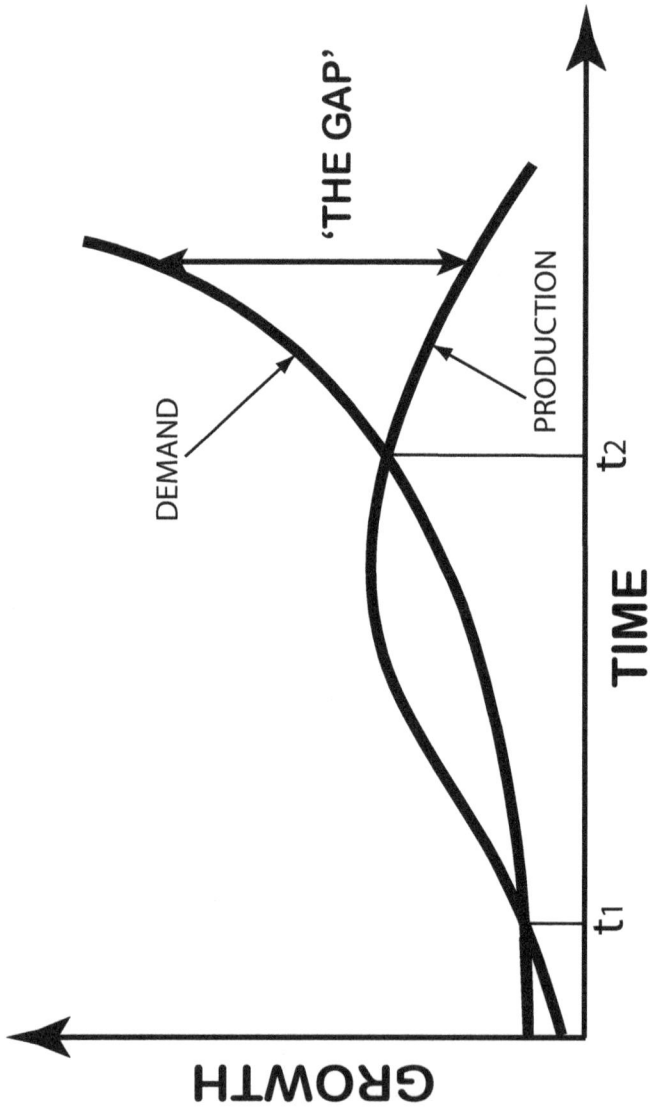

FIGURE 15.2: THE DISCREPANCY CURVE - WHEN DEMAND OUTSTRIPS SUPPLY.
It has three parts: 0 to t_1 - production (supply) lags demand; t_1 to t_2 - production, stimulated by demand, exceeds the latter (export); t_2 to ... - demand rapidly outstrips production, an ever widening gap occurs (discrepancy) which at first (but not forever) is filled by imports.

(export situation: t_1 to t_2 in Figure 15.2), and as the demand curve reaches the portion of fast growth, the situation reverses and demand quickly and increasingly outstrips production to create an ever widening gap or discrepancy. The gap may temporarily be filled by imports from areas that are still in the excess production stage. What *The Limits to Growth* suggests is this: As everybody moves in the same direction, more and more groups fall into the discrepancy stage and worldwide shortages are then inevitable.

With the uneven distribution of wealth as we see it today, it is obvious that Equation 2 fluctuates greatly from place to place. In some of the so-called underdeveloped regions, it takes the form of Equation 1 and is hardly balanced at that, whereas in the most affluent countries the demands are largely caused by expectations, the needs forming really only a minor part. Since on a global scale the demands are dictated by the affluent societies, it is important to realize that our global demand curves for all types of commodities are of a compound nature (needs *and* expectations) such as shown in Figure 15.1. Thus, growth curves are based both on need and greed. However, we should note that the demand curves in Figure 15.1 apply to the affluent western nations. For one half to three quarters of the world, growth is truly based on legitimate needs only, since in the underdeveloped countries not even the basic necessities are provided for the people, as we read daily in our papers.

It becomes apparent that the problem of sustaining man on this earth is not only related to his number per se, but also to his style of living—commonly called the standard of living. In this regard, we must note that the industrial revolution of the last 200 years has provided us with options. Our attitudes force us to exercise these options and place virtually no limits on our expectations. It is therefore evident that in order to assess our growth problems, we cannot consider population (P) alone. In some form or fashion we must make allowance for the level of expectations, or the affluence (A), of the population.

One can attempt to solve this problem at a high level of sophistication by deriving complex formulae and feeding them into a computer. This approach is neither necessary nor particularly meaningful in view of the fact that much of the required basic information is simply not available.

AUTHOR'S "NEW IDENTITY" IN THE AGE OF AFFLUENCE

We shall restrict ourselves to a very simple approach to assess the problem in its most fundamental terms. In order to establish the burden placed on the earth by a group of people, we multiply their population (P) by the affluence index (A) to arrive at the effective population (EP) in the following manner:

$$\text{EQUATION 3: } EP = P \times A$$

For the spirit of this argument, it is not necessary to develop a precise definition of the affluence index A. It is related to man's consumption of material goods such as food, energy, and non-renewable resources, such as metals. The general meaning of this index is clarified in Figure 15.3, which refers to energy consumption in various areas of the world and is drawn after Cook.[1] Since the populations in the various parts of the world are reasonably well known, it remains for us to define the numerical values for A in an acceptable manner. Arbitrarily, we shall designate the existence level which barely guarantees man's survival as:

$$A_{min} = 1$$

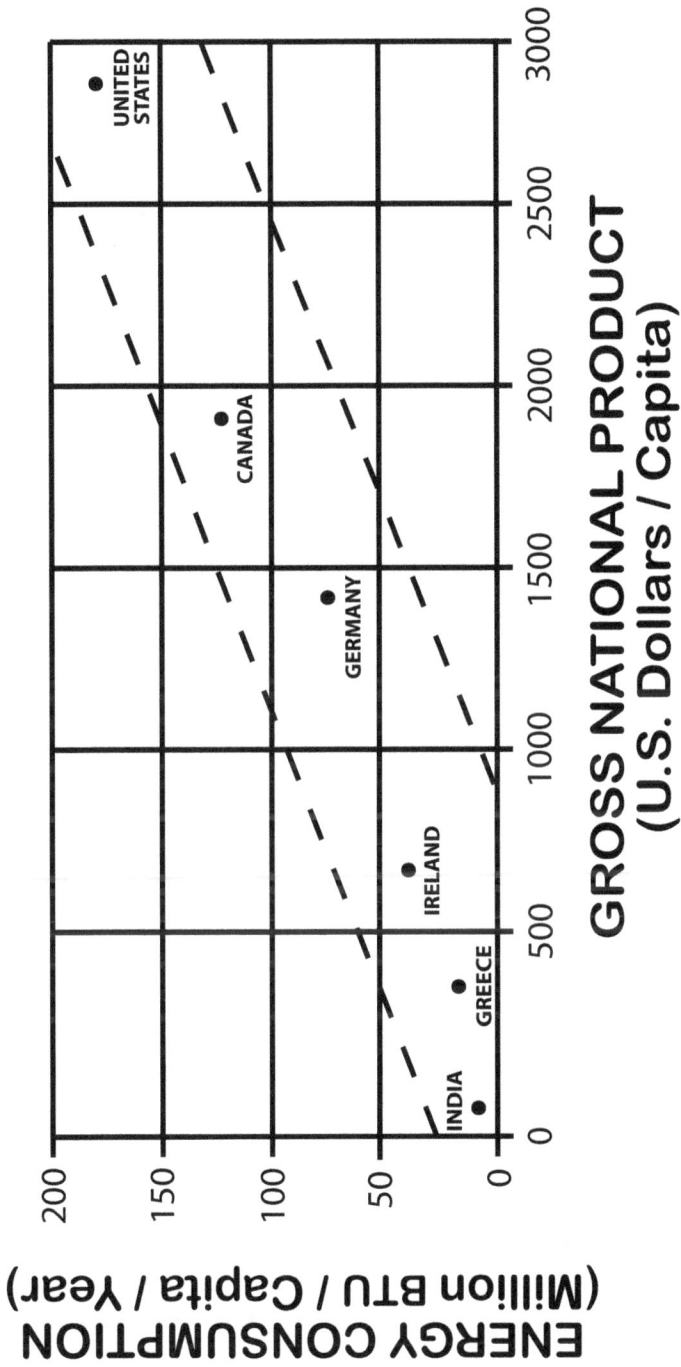

FIGURE 15.3: ENERGY CONSUMPTION VERSUS GNP (PER CAPITA).
The poorest and the richest nations differ by a factor of 20 to 50. After E. Cook, 1971.

The question then remains as to what number we assign to:

$$A_{U.S.} = A_{max} = ?$$

From Cook in Figure 15.3, as well as other sources, we can say that the following value seems to represent at least a ball-park figure:

$$A_{max} = 30$$

The total population currently stands at:

$$P_{now} = 4 \times 10^9$$

In order not to clutter our estimation with a great deal of unsupportable, detailed evidence, we make the following simplified assumption:

For $P = 1 \times 10^9$: $A = 30$ (affluent West)
For $P = 3 \times 10^9$: $A = 1$ (underdeveloped)

We overestimate the number of affluent people living at the western standard of living at 1 billion and neglect those that have standards of living somewhere in between the extremes (assigning the entire remaining 3 billion to underdeveloped countries).

We are now in a position to carry out some simple computations, which, while not accurate in detail, will show us the basic consequences of certain developments.

Our present effective population can be computed as follows:

EQUATION 4: $EP_{now} = (1 \times 10^9 \times 30) + (3 \times 10^9 \times 1) = 33 \times 10^9$

Thus, our present effective population is 33 billion. The ratio of poor to affluent people is 3:1 but in terms of effective population, the affluent outnumber the poor 10:1. Or to put it in different terms, we can make the Canada-China comparison:

$$P_{Can} = 22 \times 10^6 \qquad A_{Can} = 30 \qquad EP_{Can} = 660 \times 10^6$$
$$P_{Chi} = 660 \times 10^6 \qquad A_{Chi} = 1 \qquad EP_{Chi} = 660 \times 10^6$$

In terms of effective population, the small western nation of Canada essentially equals the largest nation on earth!

We can take these computations a step further to peek into the future. Let us assume that the present $A_{max} = 30$ is an absolute figure that cannot be exceeded (there is, of course, no indication to that effect) and let us further assume that the world population remains fixed at the present level of 4 billion (there is, again, no good reason to assume that this will be so). All that will happen in this scenario is that the total population will rise to the maximum level of affluence. The effective future population then computes as follows:

$$EP_{future} = 4 \times 10^9 \times 30 = 120 \times 10^9$$

Thus, we recognize that merely upgrading everyone to the present American standard of living will increase the effective population four-fold. In view of our unrealistically conservative assumptions (A_{max} fixed; P fixed), this is a truly frightening prospect.

Granted, this model is crude, but even so it clearly demonstrates the absurdity of an all American (or German for that matter) world. This is one of the few criticisms of Leavis[2] in which he is correct and Snow is wrong. Closing the gap will result in a reduction in the standard of living for the affluent West. All know it in their hearts, which is the only reason the topic is taboo. Just what is going to happen is a moot question, but one cannot shed the nasty feeling that fictional accounts such as *The Crash of 79*[3] come closer to the truth than many of those supposedly learned studies (research!). The only thing one can predict with some confidence: No, we are not *all* going to be rich. In fact, about the most optimistic solution we can foresee is one where all live well, but modestly (as viewed from the rich side, naturally).

The moral of this little exercise is simple—it is not sufficient to strive for zero population growth. The problem is far more involved and intimately related to our proud record as the biggest waste makers the face of the earth has ever seen. With minor exceptions, such as a few isolated tribes, everyone travels on the *road to affluence* (see Figure 15.4). As a population, through technological development, leaves the existence level it moves at increasing speed towards a higher material standard. The finite nature of the earth dictates that there be a maximum permissible such standard at which it becomes necessary to deviate from the road to affluence with a right angle turn. How do you negotiate such a turn when travelling at 90 mph? This is the basic question that all those who travel the lower parts of the road to affluence must ask themselves today. Where is the intersection and how are we going to take the right turn when the time comes?

For the affluent western nations, one cannot avoid the uneasy feeling that they have already passed the intersection and are travelling on the dead-end section of the road. If this is correct, then only a U-turn—a reduction in the standard of living—can save the situation. All indications point to the fact that such a voluntary restraint is unacceptable to western nations. But the key question—not answered at this time—is therefore: "Is this a necessity for survival?" If so, can this be brought about by backing up to the intersection, or do we have to smash into the brick wall awaiting us at the end of the cul-de-sac?

Editor's Note

While world population at the time this was written (late 1970s) was about four billion, it is closing in on seven billion today. Not quite at a 30 year doubling rate, but clearly the issues raised above are more pressing than ever.

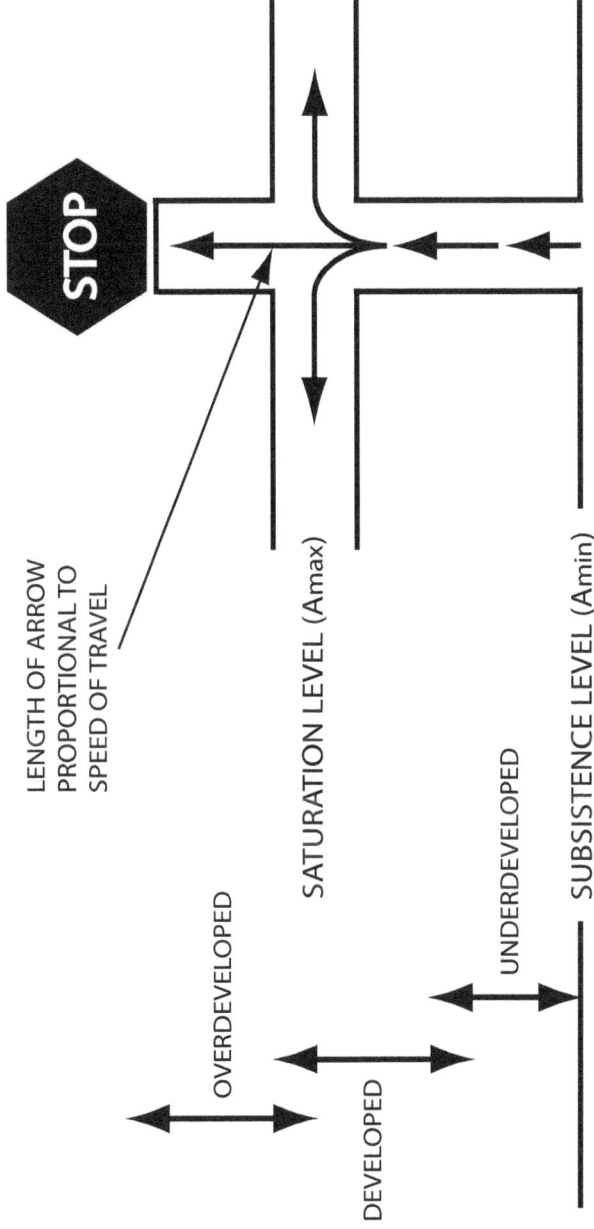

FIGURE 15.4: THE ROAD TO AFFLUENCE. Man's aspiration for material wealth lets him rise from the subsistence level at an ever increasing speed towards an excessive level of affluence. Unless one recognizes at an early stage that affluence must remain finite, it will be impossible to stop the trend once the saturation level is reached. It is difficult to negotiate right angle turns at 90 mph and even more difficult to execute U-turns at high speed!

LENGTH OF ARROW PROPORTIONAL TO SPEED OF TRAVEL

SATURATION LEVEL (Amax)

SUBSISTENCE LEVEL (Amin)

OVERDEVELOPED

DEVELOPED

UNDERDEVELOPED

STOP

16

THE POTENTIAL LEADERS:
POLITICIANS, CLERGY AND ACADEMICS

Before going into any detail on the merits and deficiencies of our potential leaders—the politicians, clergy, and academics—we can point to one thing they have in common: they are all useless. The man in the street—if there is any such animal—has received no leadership from any of these groups. His suspicions of the intellectuals are fully justified.

With the exceptions of a few mavericks, none of these groups have come up to expectations. The politicians muddle along according to the wishes of their constituents, the clergy has faded into the background of anonymity, and the academics are busy pursuing their research—read egos—in their ivory towers.

Each of these groups by virtue of their position in society should be able to grasp the problems confronting man, but none of them have made any move to alter course. It is for this reason that they have lost status and prestige. It is our right in a democracy to disagree with our leaders, but it is pathetic to see them as the target of every cartoon.

An elected clergy is as powerless as an elected political body. Spiritual leadership calls for independence and moral strength, which are just not there. In academe, it is far more acceptable to drift with the trend, both scientifically and socially, than to be a caller in the desert.

The result is that most of those from whom one could expect leadership are silent. Most of them are parents too. What will they answer their children when asked, "Dad (Mom), you know what is coming, what have you done?" Maybe none of us can point to any actual achievement. Maybe those who say that nothing can be done are right. But at least one should be able to say, "I kept kicking."

17

HUMAN VERSUS PHYSICAL RESOURCES

In the day and age of the energy crisis, we have become very much preoccupied with the importance of physical resources. Nobody can deny their importance, and human existence is unthinkable without being backed up by physical resources. We all know there are rich countries and poor countries. And, of course, in recent years we have significantly revised our thinking in that respect. Physical resources are an asset, no two ways about it. In addition, however, every group, every nation, and mankind as such, has what one normally refers to as human resources. This is an acknowledgment that man's ingenuity—his mental and physical abilities—is indeed unique in that he can overcome situations which for any other form of life would simply be fatal. History has already demonstrated this fact most forcefully in that many nations poor in natural resources have managed to live in comparative prosperity thanks to their extraordinary human resources. It seems that those nations most plentifully endowed with physical resources are the ones that place the least emphasis on their human resources, with the result that they appear to be the poor countries.

Maybe this should not surprise us. When rich in physical

resources, one can survive on that basis alone and does not have to mobilize one's human resources to any great extent. On the other hand, a lack of physical resources must be compensated for by purely human endeavour. The situation is now rapidly becoming unique in the sense that natural, non-renewable resources of all kinds are dwindling and will be scarce in a short while on a worldwide basis. Under these conditions, nobody can afford not to take a close look at human resources and lay plans for their activation. This pressing into service of man, this general mobilization, requires a change of attitude—the human revolution discussed in Chapter 47.

One can also look upon a well-nourished human being in an affluent western society as a source of physical energy. This energy, as is well known, is now usually wasted. Instead of putting it to use in constructive physical work, it is dissipated in artificial exercises, or, worse yet, it is not used at all and tends to accumulate in visible places.

It is interesting to see whether the mobilization of these latent energies can make any dent in our energy crisis. We may assume—this is only a rough estimate since the vagueness of basic data does not justify the employment of a computer—that in North America we have 150 million people capable of physical work and each one of them wastes 2,000 calories per day. You must grant that the figures chosen are reasonable or even conservative. Converting this to megawatts and comparing it to some of our power stations, we find:

Wasted human energy in North America	15 MW*
Hoover Dam, Ariz/Nev, U.S. (hydropower)	1,350 MW
Glen Canyon Dam, Ariz, U.S. (hydropower)	950 MW
Grand Coulee Dam, Wash, U.S. (hydropower)	2,000 MW
Mica Creek Dam, BC, Canada (hydropower)	2,500 MW
The Geysers, Cal, U.S. (geothermal)	500 MW

---

* In terms of oil, this corresponds to roughly 70,000 barrels per year while total annual energy consumption in North America is currently in excess of 20 billion barrels oil equivalent!

We note that even though our inactivity may be stupid, reversal of our behaviour is not going to solve any practical energy problems. Remember, also, that in a fully industrialized society, it takes about 1 MW of power capacity to supply the electrical power needs of 1,000 people. This, of course, is not their total energy consumption. Thus, the message is clear—we are hooked on energy that is far more powerful than human muscles and the mobilization of our human resources as a source of energy may be sensible and beneficial to personal health, but it is not going to alleviate what we currently perceive to be an energy shortage.

18

OBSERVATIONS ON
THE LAW AND THE LAWYERS

These observations on the law and the lawyers are made by a westerner living in a democratic country, where the laws are made by the people for the people. Admittedly, the process of changing an existing law may be a long and arduous one, but the mechanism exists. Under such conditions, the increasing defiance of the law as we observe it in all affluent western countries must be a matter of great concern. Civil disobedience is understandable in a dictatorial system, where the laws are imposed on people and where this course of action—with all its consequences—is the only way to oppose what may be considered unjust.

In the western societies, we must ask ourselves what has brought about this low public opinion of the law—this acceptance of civil disobedience. There is little question that a great deal of the responsibility must be laid at the feet of those entrusted with the upkeep of the law—the judges, the prosecutors, and defenders, or in short, the lawyers. As long as law schools are primarily occupied with teaching how to twist the letter of the law rather than uphold its spirit, we must not wonder why people begin to show disrespect.

A successful lawyer, according to our standards, is a wealthy lawyer. And just how do you become wealthy? By getting an obviously guilty person off the hook on the basis of what one usually refers to as a legal technicality. Just to cite an example: A lawyer presents a bill. The customer questions the amount and ultimately gets it reduced to one third of the initial amount. The Law Society must be informed of all such reductions that are in excess of 50%—in itself an interesting statement. It calls upon its guilty member and enters into its books a remark to the effect that this is an act unbecoming a lawyer. The particular lawyer appeals this decision to the Supreme Court, which decides that he is legally correct.

Another case. A man runs a stop sign and kills a driver on the thoroughfare. He admits that he had run this sign habitually. Result—$35 fine. Argument—all the offender did was to run a stop sign. In other words, all the bad luck is on the dead man's side.

Or an engineering society reprimands one of its members due to negligence, again an act unbecoming a professional. The court overturns this ruling, as negligence cannot be so classified (i.e. unbecoming). It did not dispute negligence as such but merely ruled it acceptable.

Or a snowmobiler runs into a barrier on a private road marked no trespassing. The court awards almost $30,000 in damages, ruling that trespassing should have been expected and the road been safe to do so.

Needless to say that the lawyers will tell me—as one recently did—to go back to my rocks, since I do not evidently understand the simplest facts about the law. This is true. I do not understand *this* law and the fact is that I have much company. And it is for this reason that we the people can no longer respect the law and must consider a judge in his garb a farce. All of us will readily admit that there are intricacies in the law which are beyond the understanding of the uninitiated, but the simple, day-to-day law dealing primarily with minor and major offences against the right of thy neighbour must be understandable to all, otherwise it ceases to exist.

For all universities this is a most serious accusation, since it suggests that something is dreadfully wrong with the law schools. Deficiencies discovered in a professional group always come home to roost where they originate, and that is in the university where such people get trained. We can no longer ignore the fact that an atmosphere of mistrust permeates the application and execution of the law in most affluent western societies. On the other hand, we cannot deny the fact that laws and regulations are needed to an ever greater degree in an overcrowded world. Advocating total personal freedom, so popular with the social scientists today, is asinine in a highly populated world. And in the absence of a high level of personal responsibility, civil laws have to fill the gap as best as possible. To suggest that we have advanced so far that such laws are no longer needed or can be violated at will, is just totally unrealistic.

Editor's Note

As a lawyer, this chapter hits close to home. I have little quarrel, however, with the basic thesis that a legal system not based on understandable concepts of justice (common sense)—and not supported by society at large—is destined for failure, or at best, circumvention.

19

ON OPTIMISTS AND PESSIMISTS

Much of the confusion that exists can be traced back to what one can label "the battle of the optimists versus the pessimists." The optimists are more pleasant to listen to, while the pessimists are usually right. This, of course, in itself represents a biased statement. The optimists are those who believe the present status can be perpetuated, that growth can continue, and that technology—a little improved perhaps—can solve anything. The pessimists are those who hold that our present way of life is doomed and drastic changes, voluntary or imposed, are inevitable.

One can give the following approximate definition of the two:

Optimist
One with unlimited confidence in technology; sees no real problem or modern predicament and therefore has no need for hope.

Pessimist
One with limited—very limited—confidence in the almighty power of technology; a person with doubts, but not necessarily without hope.

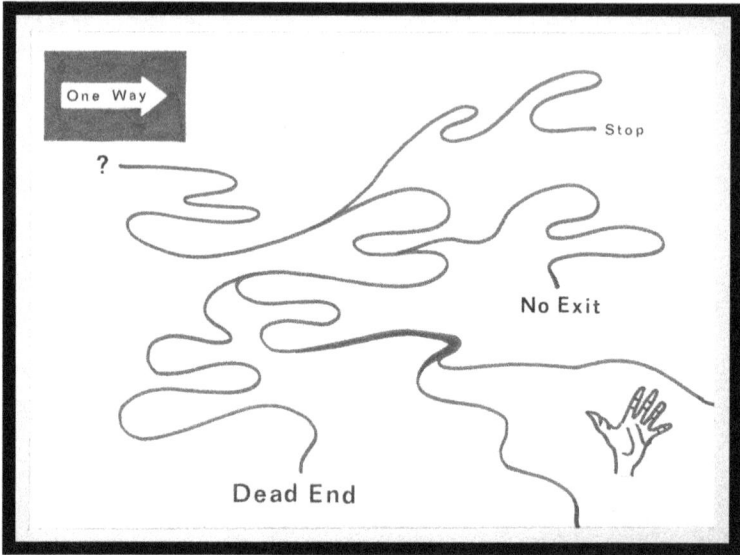

PESSIMIST'S VIEW OF "THE STREAM OF LIFE"

Pessimists tend to see themselves as realists rather than pessimists. Optimists believe that history justifies their outlook—they are wrong. The well-known phrase of history repeating itself is based on an insufficient sample. Granted, over the past few hundred or even 3,000 years, there are repetitions in the human history. However, through the industrial revolution, mankind has been moved into a new and unique situation. To deny this is not being optimistic, just foolish.

As is often the case, the positions are quite rigid, and Engel has phrased it rather well: "The optimists believe, or claim to believe, that ours is the best of all possible times in the best of all possible worlds. We the pessimists are those who fear that the optimists are right, and that man, if not the earth, is running out of time."[1]

20

THE SILENT MAJORITY

It goes without saying that in North America, the silent majority comprises the largest part of the population. At some time in the recent past, it was by many almost considered to be an honour to belong to this group, which by its very silence endorses sanity and common sense. One alluded to the fact that all the ills that befell society in the form of dissenters were really only carried by a small minority, which in due time would come to its senses and join the silent majority, or the *Establishment* if you wish.

Of course, nothing could be further from the truth. To be counted in with the silent majority must be about the worst insult that can be thrown at a thinking human being. In a time when it is painfully obvious that *the* system has had it—when it is equally obvious, except to the very youngest, that there is no place to run—in such a time to be silent and to be, in fact, an intellectual parasite, is one of the worst crimes a citizen of a free country can commit. It is equal to high treason in times of war.

This is a time of battle. Not a battle of the guns and knives, but a battle of the wits. We need all the brainpower we can get. The frontiers, i.e. the problems, are not clear and the priorities certainly not set—far less has anyone suggested any workable solutions. In such times it is not permissible for a free citizen to accept his rights without acknowledging his obligations. Any member of the silent majority should be deprived of his rights as a citizen, which, of course, would turn North America into a dictatorship, as precious few voters would be left.

Any human revolution must be preceded by an intellectual revitalization. Just because a person makes a living doing a manual job does not excuse him or her from any thinking. Since no solutions can be offered by the experts—in fact all they can provide is total confusion—it is the duty of every citizen to get involved, to assess and study the facts, and to come to his or her own conclusions. The silent majority that is never heard—that never makes up its mind on any issue—is the worst enemy of the democratic system and of mankind. Dürrenmatt is quite right when he says that a problem that concerns all must be solved by all.[1] Enforced solutions never work in the long run. Every individual has to work on the problem and once a consensus is reached, it must be carried by the conviction of all.

The silent majority must remember that the consequences of our present actions will be shared by all of us. Solzhenitsyn says they should have called out,[2] and indeed if they had, a disgrace like the Stalinist era would never have taken place. Neither would we have seen the Hitler era, nor would the U.S. have been dominated by organized crime as we see it today. Now, more than ever before, a person must be involved. One has the right to be wrong, but one does not have the right to stand aside.

Of course, the heaviest responsibility in this regard is carried by the potential leaders, i.e. the men of the cloth, the academics, and the politicians. More about their role in Chapter 16.

21

ON THE LACK OF
INTEGRITY AND COMPETENCE

Lack of integrity and competence seems to be a typical
syndrome of an affluent society. Competence appears
to become superfluous in an era where there is always
more where *this* came from. Material necessities and luxuries
are taken for granted, and no process of creation is assumed.
They are the God-given rights of man and are produced *somehow*. In today's world it is not unfair to say that, generally,
the only item on a job that exceeds expectations is the bill.
The chances that a product or a repair job will be satisfactory
run about 50/50. Needless to say, this state of affairs affects
the quality of life adversely (see Chapter 27). It is not pleasant
to constantly be *taken* as a consumer, and the only retalia-
tion seems to be to act in the same manner as a producer,
which, of course, leads to the vicious cycle in which we find
ourselves today.

The lack of competence must go undisputed. Consumer
goods today, despite all advertising and technological progress,
are often inferior to those purchased 10 or 20 years ago. This
is in accordance with the philosophy of a consumer society,
which must frown on anything called "quality." The name of

the game is consume, which in turn means products of limited lifespan. This leads to our record as waste makers, long recognized by Vance Packard.[1]

Thus, in a consumer society, a lack of competence in many areas is not only tolerable but even desirable. From a moral point of view, such attitudes are totally unacceptable. This type of production leads to an unnecessarily rapid depletion of resources, without raising the standard of living. The restoration of competence in all jobs is, therefore, one of the most urgent requirements today and the young *revolutionary* owes it to himself and the world to become proficient at some specialty. No job is so insignificant that it cannot be done either rightly or wrongly.

The lack of integrity is somewhat puzzling in a basically affluent society. It is displayed to a large extent by people who have no need for it, i.e. the affluent segment of the population. It is borne out of greed—the insatiable desire for more and more material goods and the need to have everything others have. It leads to the hectic rat race of modern western nations and usually ends with the racer being buried prematurely, a victim of his addiction. Whatever he has amassed remains behind—at least there are no known cases where someone has taken it with him.

If one divorces oneself from the rat race for a moment and takes a detached look, as much as this is possible, one immediately recognizes the futility of this endeavour. The fact is that general dishonesty is accepted today in the western world to the extent where the "upright" or honest person is simply regarded as a fool. Since no one likes to be a fool, the latter class constantly diminishes in number. To accept the violation of the rights of others in an overcrowded world can only spell doom for all. The idea that one can be a bigger and better crook than the others and therefore come out on top is childish. It leads to total social collapse with all its consequences, and many countries in the western world are precariously close to that state.

It must be recognized that the establishment of civil laws and law-enforcement agencies is not sufficient. The laws of the land must be supported by the moral principles of its people. The only real deterrent is not the uniform, but the censure of one's fellow citizens. To express one's disapproval of small and large transgressions of the law as a person to the guilty is not being a grouch, but rather a necessity. This, far more than any action by the law, will eliminate unlawful behaviour. And let us remind ourselves that in a democracy there is neither the need nor the freedom to belittle the law of the land (see Chapter 18).

Again, the lack of integrity has a detrimental influence on the quality of life. It breeds suspicion and mistrust. To love one's fellow man may be utopia and unrealistic, but one should always be able to trust a fellow man to give one a fair shake.

Editor's Note

Where do we stand today? A lack of integrity and competence lies at the root of the economic meltdown of 2008-2009, although one might well attribute the bulk of the cause to the former. After all, our captains of industry, and their brethren, the Wall Street/Bay Street *Masters of the Universe*, have been quite competent at feathering their own nests. The examples are legion, from a $50 billion ponzi scheme[2] to hundreds of millions of dollars paid in bonuses to executives who have driven their companies into the ground (the bonuses being hijacked from taxpayer bailout money).[3] We are a long way from a world where one can reliably count on a fair shake from one's fellow man.

The recent economic meltdown has caused many to reflect back on the Wall Street Crash of 1929. Ferdinand Pecora[4] comes to mind—the feisty chief counsel to the U.S. Senate Committee on Banking and Currency which, following the 1929 crash, probed Wall Street banking and stock brokerage practices. Pecora's investigation unearthed evidence of irregular practices in the financial markets by the "banksters" that benefited the rich at the expense of ordinary investors—practices driven by a lack of integrity.

The outcome of the investigation was the implementation of

regulatory controls. When a few years later, the industry was calling for a return to the "good old days" and for the measures to be lifted, Pecora's response was curt and prophetic:

> After five short years, we may now need to be reminded what Wall Street was like before Uncle Sam stationed a policeman at its corner, lest, in time to come, some attempt be made to abolish that post.[5]

Well, the policeman was called off his post[6] and we all know what happened. President Obama is calling him back.[7] We will need him to stay on that corner unless and until we reach a point where compliance with the second commandment of the human revolution (see Chapter 48) is commonplace.

22

ON THE UNIVERSAL CONSCRIPTION
OF THE MIND

Karl Schmid, teacher of German literature at the Swiss Federal Institute of Technology from 1944 to 1974, is credited with the following statement: "The universal conscription of the mind cannot be questioned." This is a wonderful way to acknowledge the fact that *Homo sapiens* represents the evolution of the brain and that it is not permissible to deny this evolution. It is also a universal statement, applicable regardless of birth, creed, religious belief, or political orientation. The superior brain is the unifying bond of mankind. To refuse to use it to the fullest extent is a sin by any religious standard and a fatal mistake regardless of race or class. Karl Schmid has formulated in a most precise way one of the natural laws to which man is subject. To violate it is to court disaster.

It also confirms that the great minds in the humanities and the sciences reach identical conclusions, even though they approach things from very different angles. The above statement has general applicability and is not affected by the fragmentation of knowledge as we see it today. It could not have been made by one who religiously believed in, and adhered to, this fragmentation. It also directs itself against those that feel as narrow specialists they are doing their share. Quite obviously, to be an expert in a narrow field does not acknowledge the above requirement—it does not constitute using the mind to the fullest extent as demanded above.

While technologists may appreciate the elegance and precision of the above statement, it takes a humanist to formulate it in such terse form.

23

COMMUNISM AND CAPITALISM
EQUALLY DOOMED

First, let us be clear that we are discussing communism as known to us in Russia, particularly through the writings of Solzhenitsyn. China is obviously an anomalous case and while it is counted amongst communist countries in general, it cannot be considered the same thing in this chapter. It is important to compare the two systems (communism versus capitalism), since many people believe that our current predicament could be alleviated by a change in political philosophy.

Toynbee feels that in order to survive, we must at least have a temporary dictatorial system in the West.[1] This contrasts with Lorenz's view, where he states that a visitor from outer space would find little difference between communist Russia and the democratic West.[2] Many people, though, and among them some outstanding thinkers, seem to agree with Toynbee that the democratic system has led to excessive and misused freedom of the individual, and that corrective action such as the human revolution (see Chapter 47) requires the establishment of a more authoritarian form of government.

We should throw a word of caution into this discussion.

Present conditions clearly indicate that the mere establishment of a dictatorial form of government in the West is no guarantee of the elimination of our predicament. Reading Solzhenitsyn, and in particular his short stories, as well as from other sources, we find that crime is also rampant in Russia. Pollution is certainly not absent and has not been given any more priority there than in the West. The citizens are committed to an ever increasing standard of living, even though they may presently still be on the lower segment of the road to affluence (see Figure 15.4). Thus, to simply replace democracy with tyranny is no assurance whatsoever that our real problems will be solved. We may end up with all the well-known disadvantages of a dictatorship—and the examples of these are before our very eyes—and yet still not be free of what troubles us presently. In fact, one may very well argue that dictatorships *must* be committed to the road of affluence in order to stay in power (Hitler!).

To dictate the necessary measures of self-denial in a material sense from up on high will be just as unsuccessful in an authoritarian as in a democratic system. In order to bring about real change, it is necessary to reach the individual and to gain his voluntary cooperation borne out of a sense of understanding. Whether or not this is what the Chinese are trying, and whether or not they have succeeded in some measure, it is too early to say.

I would oppose the view, particularly prevalent in some universities, that to abolish the capitalist system and become a *left-winger* will solve anything. We have every indication that the left is as greedy and egocentric as the right, and there is really not much to choose between the two. I would certainly agree with Lorenz's view. Does it matter whether your title is vice president and you own property on Vancouver Island, or you are a commissar and own a Dacha at the Black Sea? I fail to recognize a fundamental difference. In Russia, the secret police is making life plain hell for a great number of people, and in the West, organized crime operates in a somewhat subtler way but still makes life very unpleasant, to put it mildly, for a possibly lesser number of people. In both cases, those that are not directly stepped on look the other way and pretend that all is well.

I am afraid that even a cursory assessment of the situation clearly demands that one look for a *new* solution, and that, of course, is a hell of a lot more difficult. It is far easier to compare existing and past systems, and assess their respective flaws and merits, than to produce something brand new. This requires originality—a quality suppressed in *Homo technicus*. Yet our previous record shows that so far nothing has worked. Now we are in a position where it either works or we are out—clearly, a third down (fourth down if you live in the U.S.) and long situation.

24

ON MATERIAL AND
INTELLECTUAL PARASITES

A material parasite is obviously one who consumes a great deal more than he produces. We all know that there is no shortage of such individuals. Their number is carried by those who contribute more than they ask for. On the whole, and in the long run, this equation must balance. Should the percentage of material parasites exceed a critical number, then obviously the system can no longer function in an economic sense.

In this chapter we are, however, far more concerned with intellectual parasites, those that believe they can go through life denying man's heritage—the evolution of the brain. It is not sufficient to have an intellectual elite comprising a very few percent of the population, while the rest do not care—do not make an attempt to understand, discuss, and dissent with the concepts evolved. This means that the intellectual forces of humanity operate in a vacuum and have essentially no effect on the destiny of mankind. For philosophers to live in a world by themselves—neither heard nor understood by 95 percent of their fellow men—is a totally futile endeavour. And that is, of course, exactly what we have. To feel that because one earns a living in a manual job one is exempt from any intellectual gymnastics is wrong—just as wrong as those who do no physical work and feel any physical exercise is the exclusive prerogative of intellectual morons. Many schools carry the logo: *Mensa Sana in Corpore Sano* (a healthy mind in a healthy body), clearly demonstrating that one does not exclude the other.

We show elsewhere (see Chapter 36) that modern technology has done little to alleviate this situation. In fact, everything points to it getting worse. Laziness is an attribute of modern affluent societies. The evidence is seen everywhere. The point is that physical laziness is easily visible and demonstrable while intellectual laziness is far more subtle and less obvious, but at the same time its consequences are far more devastating.

25

CHINA: A CLOUD OR A RAY OF HOPE?

It is very difficult for most of us to obtain a clear picture of the Republic of China. Reports by those who have had the opportunity to visit are all biased either in favour of, or against, the current Chinese experiment. Many actions must arouse one's suspicion. The Chinese have wiped out Tibet, the small nation on the roof of the earth. They support communist movements in various parts of the world, which is contrary to their assurance that they are a peace loving nation concerned only with the well-being of China.

One fact seems to emerge from all the reports of those who know the old as well as the new China. There has been a tremendous change. Nobody is hungry today, ordinary crime is virtually nonexistent, and the current government believes that this different lifestyle has to be based on individual attitude. The good of the individual comes second, and the good of the group and the nation takes precedent. The individual is part of the group and only part of the group. This holds until you reach the very top, where the Chairman resides in solitude as the ultimate source of wisdom. And again you ask yourself: "Is this for real, or are we being taken in?" You look at the little red book and discover much wisdom, but also the fact that no individual human being is infallible. Just to cite one example, Mao states: "For the purpose of attaining freedom in the world of nature, man must use natural science to understand, conquer and change nature and thus attain freedom from nature."[1]

This is, of course, exactly our downfall in the West—our obsession with conquering nature, to view it as if it were created for the express purpose of serving us and our desire for a hedonistic life. Maybe Harry Schwartz put it in a nutshell when he said that in order to cope with us they (China) had to adopt our ways.[2] Maybe that is all that is to be said about the Chinese experiment. The growing volume of Chinese industrial products in western markets certainly points in that direction.

It is interesting that anthropologists, when discussing the variability of human nature, always refer to isolated, small tribes and never to the Chinese, who represent one fifth of the world's population. Just what has happened to aggression in China? Is it merely suppressed by outward direction?

26

SWITZERLAND AND NORTH AMERICA

In Switzerland, the democratic process is almost 700 years old, while in North America it dates back 100 and 200 years. Switzerland is of a size that requires a magnifying glass to locate it on any small-scale European map, while Canada and the United States occupy a continent. Evidently, any comparison between areas so different in time and space is dangerous. However, almost no author can refrain from drawing on his previous experience—and this one is no exception.

One interesting question one might ask is: "What is the reason for the stability of Switzerland?" Obviously, there is more than one. All Swiss know they are better off as Swiss than as part of Germany, Italy, France, or Austria. This may well be one of the most powerful arguments. However, the Swiss have taken a very real interest in the democratic process. Even to this day, problems of the country are hotly debated in every pub. It is no coincidence that it was the Swiss writer Friedrich Dürrenmatt who said, "What concerns everyone can only be resolved by everyone."[1] While it must remain an unproven statement, in my mind this concern of everyone—not just the intelligentsia—with political problems is one of the main reasons for the stability of Switzerland. It is this acceptance of individual responsibility that accounts for the continued existence of the state.

That this state of affairs is threatened and undermined by the acquisition of affluence and the rise of the mass media, such as television, seems to be a fact. After all, it was one of my Swiss friends, Paul Eggenberg,[2] who coined the term "affluence anarchy."

27

SOME REFLECTIONS ON
THE QUALITY OF LIFE

The expression "Quality of Life" is much used recently, both in publications and speeches. It seems, however, that nobody has ever attempted to define the term. It may well turn out to be indefinable, having different meanings for different people. This should not deter one from attempting a definition. One thing is clear, that many authors simply equate the quality of life to the standard of living:

EQUATION 1: QOL = SOL

The above equation would imply that the higher the standard of living, the better the quality of life. No one will deny that there is a certain relationship between the two, but it is not likely as simple as indicated above. Figure 27.1 shows an attempt to relate the two terms in a more realistic fashion. It is clear that a standard of living near the existence level is likely to be related to a low quality of life. A life full of worries, disease, and hunger is not likely to be very dignified. As the standard of living rises, so does the quality of life. At first, this rise is rapid. As the basic needs are fully met and some of the more reasonable expectations (see Chapter 15) realized, the rise in

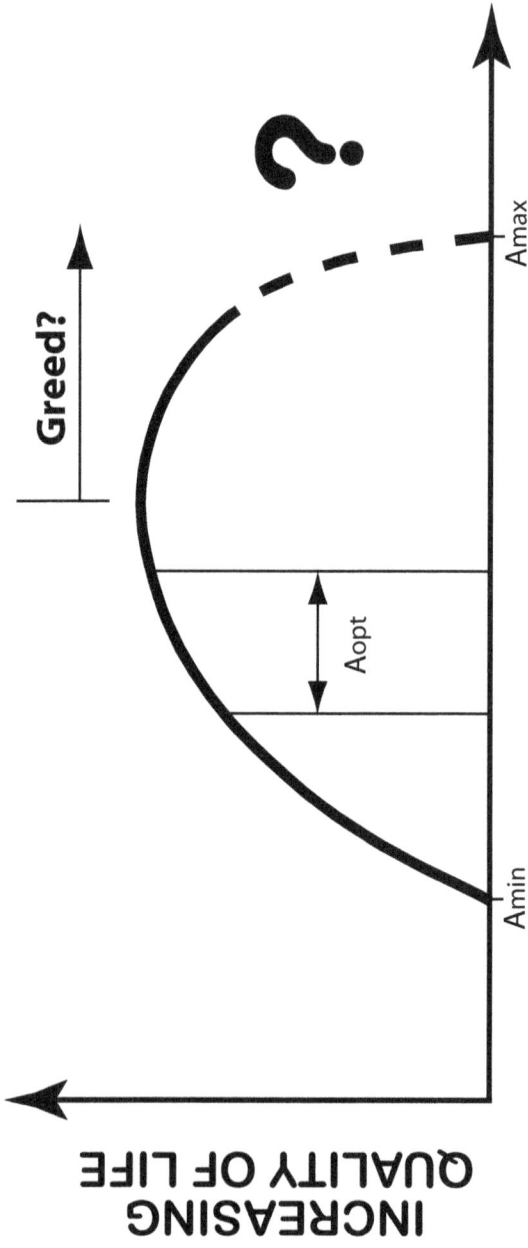

FIGURE 27.1: RELATIONSHIP BETWEEN QUALITY OF LIFE AND STANDARD OF LIVING.
At the subsistence level, the QOL is non-existent. As affluence rises, there is at first a rapid, followed by a more gradual, rise in the QOL. As affluence passes a critical value, the QOL begins to deteriorate and may well be reduced to zero. A range of optimum affluence (Aopt) is shown. For this range, a high QOL calls for a minimum taxation of life supporting systems.

the quality of life begins to flatten. As the standard of living reaches a point of saturation, where all material wishes are fulfilled, boredom sets in (lack of stimulation), no goals seem to be left, nothing to be achieved, and the result is a rise in crime and the hectic race from nowhere to nowhere which we see in our affluent areas today. The feeling of total superiority spreads, personal freedom is the call of the day, and *affluence anarchy* sets in. Personal freedom is all important, and the rights of others are no longer respected. In a society basically free of hunger and cold, petty thievery and other violations of the law are rampant (you need a six pound chain to secure a three pound bike).

All indications point to the fact that as far as the quality of life is related to the standard of living, most areas today are not well situated. The underdeveloped countries sit far to the left of the peak, while the affluent societies are obviously already to the right of the peak and enjoy a state of affluence that is detrimental to an optimum quality of life.

It is also evident that the quality of life is not uniquely related to the standard of living. Trust in your fellow man is an important agent in gauging the quality of life. Obviously, both excessive poverty and excessive wealth are detrimental to that concept. However, it can exist and be independent, to some degree, of the standard of living. It is again a function of our values and moral standards, which are affected by the standard of living but need not parallel the latter rigidly. Thus, a high quality of life demands the development of human values different from those currently existing in most, or almost all, societies.

This again calls for the human revolution (see Chapter 47)—a change in attitudes and values. It also requires technological efforts, since a minimum standard of living must be created for many before it is possible to have a reasonable quality of life.

28

TRAINING VERSUS EDUCATION

G enerally, we are used to referring to our school systems as educational facilities, when in fact they are
mere training centres. It is important that we learn
to make this distinction, particularly since in one field we
are successful, while in the other we are not and never have
been. The high standards of technology in many fields around
the world clearly indicate that in terms of training we are not
at a loss. Certain deficiencies that have appeared are due to
attitudes, rather than proper preparation for the job. Thus,
on the whole, we can have little quarrel with our schools in
terms of training.

One word of caution, though. The present state of the
world is attributable to those who went to school some years
ago—in times when discipline and the three Rs were still in
the teacher's vocabulary. That this has all changed recently
is well-known, and whether the current training is as effective
as that of 10 or 20 years ago is open to question. Be that as it
may, the training aspect of schooling can be relatively easily
understood and assessed.

It is a different matter with education. This process, by
my definition, means to learn to understand the world, our

place in the world, and, above all, to get to know one's own self, which is equivalent to the acquisition of wisdom. In this regard, our modern history—the last 10,000 years—is a dismal failure in that we have never educated anyone.

It is very questionable whether one can in fact educate another person. The best one can do is stimulate the natural intellectual curiosity—which should exist in every human being—and encourage the personal commitment to intellectual activity that is truly the mark of any so-called educated person.

History demonstrates that we have had no success in this regard. Just look at the Greek play and what do you say: "Marvellous, how this author 2,000 or more years ago captures human nature!" Of course, you have just admitted that your neighbours, after 2,000 years, are no different than the Greek characters that just walked across the stage. A clear indication of zero change. There is no great hope that this will improve as long as we insist on serving up our knowledge in little bits and pieces to the next generation. The fragmentation of knowledge as it exists today (see Chapter 12), and has for a long time, has a direct bearing on the problem.

To educate people is to bring about the human revolution (see Chapter 47).

29

WHAT MAKES A UNIVERSITY: DIPLOMA OR BACHELOR?

One of the dilemmas of any modern university is its dual commitment to training and education.[1] Today, no university can afford to be only an ivory tower. And yet, to banish the ivory tower atmosphere from a university is to destroy its very essence. Solutions to such problems, if they can be called solutions, usually take the form of compromises. However, before examining the present situation in detail and exploring an alternative to the current state of affairs, it is necessary to revisit the distinction between the terms training and education, as discussed in Chapter 28.

Training results in competence. It is concerned with the transmission and application of knowledge. Due to the enormous range of man's collective knowledge, training is always specialized and never universal. The acquisition of knowledge through training permits a young person to take his or her place in society and to make what is hopefully a useful contribution to that society in terms of its continued existence. Training, thus, allows one to become a producer.

In return, such persons will be given an appropriate remuneration (at least in theory) which permits them to become consumers as well and share in the created material wealth. This description obviously applies to anyone with special skills. Thus, training is not the exclusive prerogative of the university graduate.

It is fair to point out that as we proceed from the apprenticeship and vocational schools to the technical schools and colleges and eventually to the universities, training becomes an increasingly prolonged and more intensely intellectual exercise, and dexterity, so important in the trades, takes a back seat.

It is clear, however, that on the basis of training alone, the universities cannot lay claim to a unique position; yet they *do* claim such an exceptional status.

Education is a far more elusive term then training. It is evident that I do not refer to the common usage that, in fact, often equates education and training. The process of education reaches far beyond that of mere training. As noted in Chapter 28, it is even questionable whether education can be passed on in the same manner as knowledge. It seems far more likely that "educated" refers to a state of mind—a commitment on the part of those so designated.

If this is, indeed, the case, then the best any generation can do is to generate a climate that encourages successive generations to make that commitment. One of the outstanding features of an educated person is a wide-ranging, if not universal, intellectual curiosity—a curiosity that acknowledges the world is a whole and must be considered in this context. Clearly, this is almost a direct contrast to the goal of training, where specialization is a necessity.

Such a broad view of our surroundings inevitably forces one to come to terms with one's own position in the universe. Such contemplations, by necessity, lead to the adoption of ethical principles, which, in turn, prescribe a moral code.

Thus, a truly educated person also has *integrity*. Observation of the world around us makes it abundantly clear that knowledge without integrity is not only useless, but outright destructive. Integrity, contrary to knowledge, is not divisible. One can be knowledgeable in only one field, or one can have only limited knowledge of a field, but there is no such thing as partial or limited integrity. As such, the use of the phrase "intellectual" honesty must be rejected as erroneous. A person either is or is not honest.

Education thus emerges as an integrated concept, in contrast to the specialized notion of training. Surely, the full definition of an educated person is a difficult one and is open to question. Possibly the best we can do here is to describe him or her as a knowledgeable, widely curious person with strong moral convictions. We normally refer to such a person as *wise*.

Many universities give lip service to this dual commitment. A faculty of arts and sciences, for example, may have three departments: the humanities, the social sciences, and the natural sciences. A student majoring in one of the departments might be required to complete two full courses in each of the other two in order to graduate.

This is our contribution to the education of the coming generation. Engineers, similarly, are asked to only complete a few outside options in order to graduate. That the concept does not work is amply demonstrated by the thousands of university graduates in North America whose interest in anything beyond their specialty and sports is virtually nil.

Universities are not unique because of the often repeated cliché of "teaching and research" (maybe I got this one reversed?), but rather because of their role to educate as well as to train, or at least to create a climate favourable to self-education. The fact that presently universities are successful in one but not the other does, indeed, deprive them of their unique position.

One of the basic reasons why mankind finds itself in a predicament today stems from the failure to unite knowledge. There is a general reluctance and aversion to view the whole and a great preference for intense study of special fields without ever putting them into a broader context. Any attempt to create a favourable climate for education must recognize this fact. C.P. Snow's concept of the two cultures is as valid today as when it was written in 1959. However, its implications are far more frightening today than they were at the time of writing.

For technologists and humanists to live in mutual ignorance is not only stupid, but immoral and hard to reconcile

with the high opinion both groups have of themselves. Therefore, any alternative to the present system must attempt to
alleviate this situation. The following is suggested:

1. Universities recognize that currently their degrees
   are professional only and not university degrees in
   the true sense of the word.

2. Universities create the following option, open to all
   students intending to obtain professional training—
   a one year course structured as follows:

   i.   Students are divided into groups of 25.
        Each group is looked after by five professors
        representing: engineering, natural sciences,
        humanities, social sciences, and the fine arts
        (man is also unique in that he has a sense
        of beauty).

   ii. Each team of professors looks after two
        groups of students for the year. The nor-
        mal teaching load of those leading the pro-
        gram will be drastically reduced or dropped
        entirely. Since many service courses will
        vanish under the new program, this proposal
        will not demand as much extra manpower as
        might be envisaged.

   iii. Courses will mostly take the form of seminars
        and discussions. There will be no examina-
        tions. Students will earn credit by oral and
        written contributions. This will be a pass-fail
        exercise and the verdict will be based on the
        student's commitment to the course. Profes-
        sors will be present in the audience as often
        as possible and will not just act as lecturers
        and discussion leaders.

iv. Topics covered will vary according to the composition of the professorial team. However,
in all instances, the world will be treated as
a whole and man as a part of it. To ensure
exposure to varied viewpoints, student groups
will change professorial teams mid-year.

v. Students can take this course either prior
to their professional training, interrupt such
training for one year, or take the course upon
completion of their professional degree. Passing this one year course will entitle the student to the degree *Bachelor of...* whenever the
appropriate professional program has been
completed.

In this manner, the student acquires a true "university"
degree. This alternative provides for a total immersion of the
student for one year in all aspects of man's history, his cultures, his surroundings, and his technical achievements during his rapid rise in the recent geological past. The program
should not only contemplate the past, but an attempt should
be made to glimpse into the future, even though some of our
colleagues may not consider this to be a scholarly activity.

Service on the professional teams will open up a new world
for many a colleague. The one (or more) year tour of duty will
serve better than any sabbatical to renew vigour and permit
many to return to their own specialty with a new sense of its
significance in the greater world.

30

THE WORKING MOTHER: A MYTH

We have already concerned ourselves several times with the unique aspects of man. He is also unique in that the rearing time for his young is very much longer than that of any other species. Since the learning process in man is so advanced, this is not surprising. Even though part of the rearing is done in institutions (schools) and by society at large (peer group), much of it remains the responsibility of the parents. Psychologists have also come to realize that the early childhood years are the formative years. During these years, the influence of the mother is the predominant one, and her role cannot be emphasized enough.

The term *working mother* seems to imply that a mother is a member of the workforce. We all, however, know that this is not so and to the contrary; the implication is that being a mother (and housewife) is no work at all. A *working mother* is a mother that does something "useful"—read making money—besides rearing children.

Never have we invented a more asinine concept that has had more devastating consequences. Being a mother and house-wife is a tough and frustrating job and a full-time job at that. That women wish to get out of the house more and away from

155

it all is understandable. But to take a full-time job while you have three children between the ages of 2 and 10 at home is the utmost in abdicating your personal responsibilities. We hear such remarks as: "Isn't she wonderful, four children at home and the professor besides, and she does such a wonderful job of it." Only the uncritical North American mind could invent such a stupid, Dale Carnegie inspired statement. The fact is that this woman is almost bound to do a lousy job of both, simply because no human being carries two full-time jobs successfully.

The proof is simple. Just look around at how many middle-class marriages split up and how many wealthy to semi-wealthy kids go wrong. Women chasing the mighty buck carry a heavy responsibility for our failure to raise the next generation. Certainly, it would be unfair for us men to heap all the responsibility onto the women. We have much to answer for ourselves, but that should not detract from the fact that the glorification of the working mother is both wrong and immoral.

This is not to say that women should be a downtrodden and discriminated half of the human race. But liberation cannot be bought by the abdication of what is a naturally imposed, personal responsibility. As Simpson has stated: "...personal responsibility is non-delegable."[1] The daycare centre is *not* the answer. Particularly when we see people screaming about a rise in fees from, say, $70-$80 per month, while at the same time they board a horse for $120 per month. Just where is our sense of proportion?

Let us also not forget that there are women who must work, forced into this situation by circumstances beyond their control. But to just hold a job because you are not the mother or housewife *type* is criminal. If this is your attitude, then have no offspring, but to have your cake and eat it in this case is going to give you more than indigestion.

Thanks to modern medical technology, it is no longer necessary to keep all women barefoot and pregnant at all times in order to ensure the survival of the species. Women do, indeed, have other options today. But like everyone else, they cannot violate the natural law that says you can't have your cake and eat it too. To have a number of young children and to discover that one is not cut out to be a mother and delegate what is possibly the foremost personal responsibility to a day care centre—this is the greatest sin one can commit against mankind.

Editors Note

Wow. Certainly not politically correct, but it reflects the times in which it was written. I would say the basic premise remains sound today—full-time childrearing (in the formative years at least) is not something we should, or can afford to, institutionalize (read daycare). Where I would differ somewhat is the apparent view that this full-time childrearing responsibility belongs exclusively to the mother. Either parent (Mom or Dad) can take on this role. The important point being that the child enjoys the full time attention of one of its parents in its formative years.

Apparently, this is still not politically correct. When Alberta's finance minister suggested that good parenting means "...you don't both go off to work and leave them for somebody else to raise...," the attack was fast and furious: "These are truly outrageous claims. I have never been as stunned by the sheer arrogance and ignorance...as I am today. In a sense, Iris Evans did us all a favour by revealing her contempt for the sacrifices made by hardworking Alberta families."[2] Foregoing income and having one parent devoted to parenting is apparently not "hard work", nor any "sacrifice" worth recognizing. If children are the future (hard to argue with that), why do we place such a low priority on parenting?

Having children—a right or a privilege? Two recent items in the news highlight the ethical questions involved when the technologists or toolmakers forge ahead without restraint.

The first story involves "Octo Mom"—a single mother in her mid 30s living with her parents and her six children, including a set of twins, between the ages of two and seven. These children had been conceived through in vitro procedures in which six embryos were implanted. Apparently, six was not enough.

She pursued further in vitro procedures, again implanting six embryos, resulting in nine-week premature octuplets (two of the embryos led to twins) delivered by Caesarean section, for a total of fourteen children.[3] To their credit, some of the technologists agreed that something was amiss—"If a medical practitioner had anything to do with it, there's some degree of inappropriate medical therapy there."[4]

The second story concerns a 60 year old woman who was denied in vitro fertilization in Canada due to her advanced age and proceeded to have the procedure performed in India[5]— "I had my heart set on it. I wanted a baby." She returned to Canada for the birth, which resulted in twins, born seven weeks premature by emergency Caesarean section after severe hemorrhaging and following four weeks in hospital. The birth of the twins was heralded by another mother—"Sixty's just a number"—who also had twins when she was 59, following in vitro fertilization.[6] Her reason for having children so late in life—she said it would make her happy.[7] It was reported that she was defending people's right and prerogative to have children at whatever age they wished[8]—perhaps not a surprising sentiment from a member of the *me* generation.

What is wrong with this picture? What about the best interests of the child and of society? Having children is not an automatic right. There is a responsibility to ensure that the children brought into this world will have a decent childhood, which, after all, is the launch-pad for life. In the one case, the child has 13 siblings and will be raised by a single parent. In the other case, the child has parents who may well die before it graduates high school. In both cases, the child is born significantly premature and society's health care system is severely taxed. Neither case provides the child with a fair shot at a healthy and rounded upbringing and life.

Further evidence of the technologists running ahead of the humanists— *Let's do it if we can*—with little thought for the consequences of their actions.

31

THE TIME OF LEISURE: ANOTHER MYTH

Much is said and written today about our excessive leisure time and how to spend it. This, of course, is a myth. While none of us can forecast the future accurately, the probability seems to be for more and harder work. We come to realize today that in the western world we are presently enjoying a standard of living which we do not deserve; in fact, we have mortgaged the future with our behaviour. The North American continent is a mess, from the Gulf of Mexico to the Arctic Coast. All in the name of economy, which is nothing more than a synonym for laziness. We go in, rape the country and leave behind a mess that will somehow take care of itself.

The conservationists want us to stay out. But this is obviously not the answer. To feed and clothe and comfort the almost 250 million people in North America, we have no choice but to exploit nature. However, we can do it on an economic (least cost) basis only, or in a responsible but not so economical way. The latter requires that for each "productive" worker, there are a number of unproductive workers addressing themselves to the cleanup job. This simply means more work (expense) for the same output.

159

Upholding the law is another non-economic proposition. However, it is a necessity in order for a society to survive. The ever increasing disrespect for law and order indicates that amongst other factors, there are not enough people looking after the law—policemen, judges, and lawyers.

In Canada, we presently act as if world government were a reality. Our army is essentially nonexistent. How do we expect to enforce a 200 mile fishing limit? On three oceans? With a number of Newfoundland fishing dories manned with crews carrying sandwich boards reading "No Trespassing Please"? In our present world, no country that wants to preserve its way of life can exist without any armed forces. Soldiers have never been known for productive work. So here, again, we will have to support more people in a necessary but unproductive function.

In a country where people currently are too lazy to push their shopping carts back to the store entrance, where garbage is still customarily thrown into the ditch, where our most "educated" leave their student centres such a mess that it makes your hair stand on end—in such a society, the concept of eternal leisure time is a dream that remains a long way from reality.

32

CONSUMERS VERSUS PRODUCERS

A very curious distinction is made between consumers and producers in most societies, implying that one is a totally different type of fish from the other. The fact that basically everybody is both a producer—of some kind or another—and a consumer seems seldom to surface. Better known is the dichotomy that the consumer wants to get the most for the least—which is as old as man, presumably—and more recently, at least in western philosophy, the producer wants to produce the least for the most. The strikes, absenteeism, extended coffee breaks, etc. provide ample evidence for the latter statement.

To offer the least for the most and ask the most for the least leads to an imbalance which is obvious. One does not need a university level course in economics to realize that such a situation cannot be perpetuated ad infinitum. The riches of a blessed land such as Canada may permit such insanity to occur and exist for a limited time, but in the long-run the bill must be paid. This attitude is equivalent to the postulate of a *perpetuum mobile* (perpetual motion) in physics, which would violate the laws of nature as we know them.

It is, in itself, astonishing that the producer attitude of the least for the most—a *rip-off* in plain language—can develop in a modern society where interdependence is so complete. One would expect that even the slightest reflection by a producer would lead to the realization that since he is also a consumer, he himself will be the first victim of such an attitude. Again, such behaviour does not enhance trust in one's fellow man and decreases the quality of life (see Chapter 27). It is contrary to the concept of affording your fellow man a fair shake and, as such, is bad since it does not enhance the survival of the species.

It is about time to recognize that the lot of the plagued consumer can only be improved by adjusting his own attitude as a producer. To provide a decent job for decent pay is not an old-fashioned and outmoded work ethic, it is simply a common sense recognition of what is required to keep things going.

33

RIP-OFF:
A SYNDROME OF AFFLUENCE ANARCHY

Excessive wealth leads to social decay, a condition where no one cares for anything except his or her own well-being. This state is typical in all affluent societies and varies only by degrees. The concept of a *rip-off* is usually confined to large (faceless) corporations, preferably multinational ones. However, without the benefit of a university course on the subject, day-to-day life demonstrates beyond the shadow of a doubt that this condition permeates all segments of society. The only thing that varies is the degree of success, some groups being more skilled than others. Given the opportunity, everybody rips off everybody. When the job is done, the only part that exceeds the expectations is the bill (see Chapter 21), which is often padded to the extent where it borders, or even transgresses, the legally permissible.

The food store ethics, so well exposed many years ago by Vance Packard, deserve special mention.[1] Peas—19 cents a can; "Special" — 5 cans for a dollar. Or have you ever bought a pack of detergent for its regular price? The examples are legion.

This philosophy again has its origin in a world that believes only in material wellbeing as the ultimate good. That many people in the world live at a level where material improvement is the only goal is understandable. When such improvement means survival, or at least a dignified life at a modest level, it is not only understandable—it is mandatory. But in the affluent western world, such striving is pure greed. The fact that the urge for material possessions is insatiable in man will prove to be his downfall. The invention of property in itself is not bad, what is bad is the fact that there is no ceiling.

34

REFLECTIONS ON THE GENERATION GAP

The generation gap is not a new invention, although every generation seems to think so. The question that poses itself today, and undoubtedly has been posed before, is this: "Is the generation gap as we see it today greater than ever before?"[1] No doubt that almost every generation before us was willing to answer the above in the affirmative—the problem being that each of us lives only once. The fact that the human situation today is somewhat unique in man's history, permits us to ask this question with possibly more justification than in the previous 1,000 years.

Before we can address ourselves to this problem, we must ask what causes the generation gap. Two factors are at work: (i) that a young person looks at the world differently than the middle-aged or older person; and (ii) that the world we live in is subject to continuous change, largely inflicted by man himself these days. The first of these factors has been at work throughout man's history, and there is no reason to believe that today is any different than any other time. The second factor is also one of continuity. However, in this case we not only have reason to believe, we in fact know, that the rate of this change has been greatly accelerated during the

20th century and, in particular, since World War II. There can be no question, therefore, that a young person today experiences a world which is vastly different than the one seen by the previous generation in its youth. This simple consideration permits us to confirm our suspicions that the generation gap today must, indeed, be larger than ever before in man's history. In this regard, we find ourselves in agreement with Lorenz.[2]

This, together with the preponderance of youth, has far-reaching implications. Man's continued existence depends on the transmission of cultural values from generation to generation. This handing down of accumulated human wisdom has never occurred problem free—with youth in its role questioning the traditional values, sometimes rightly, sometimes wrongly. It is customary for youth to break the traditions, whose values are not necessarily apparent at the age of 20—an age that seems to always be willing to try new ways and abolish old. It is equally customary to find youthful revolutionaries mellow with age, recognizing that what seemed a nuisance is in fact a necessity. A large generation gap, with disrupted communication lines, together with a numerical superiority of youth, makes the transmission of values an exceedingly difficult task today. One cannot help but feel that in the last 20 years, the western societies have increasingly failed in this role.

35

ON FOUR LETTER WORDS

Recently, four letter words have become respectable in the literature. Fortunately, the English language being simple, this does not result in a great enlargement of the vocabulary. In fact, knowing a single word will be sufficient to demonstrate that you are *with it*. What is lacking in variety is made up by frequent use as noun and adjective. What has gone almost unnoticed is the fact that some four letter words are so dirty that they have been eliminated from the vocabulary, particularly by our younger generation. The word in question that has gone out of style is WORK.

36

THE MOST DEVASTATING INVENTION

Asked what they consider the most potentially dangerous invention of modern man, most people would, without hesitation, put their finger on the development of nuclear energy. There can be little doubt that the misuse of atomic power can have devastating effects. However, the most dangerous invention ever devised is of a far more subtle character. It is one of these many cases where technologists have been carried away by their curiosity and the challenge of the technically feasible, without ever asking the question whether the invention should be made at all. I am referring to television, an instrument that destroys our most valuable assets of creativity and originality

Before we go any further, let me be clear that I am not referring to the type of programming broadcast in most parts of our civilized world; I am referring to television per se. Watching television keeps the viewer entirely passive. There is no room left for his imagination, as in reading or radio (the "theatre of the mind")—listening to the spoken word.

Information—usually irrelevant though this could be changed—is showered on the victim at a relentless speed, far in excess of the absorption capacity of the human mind. It does not help to improve the *transmission* when the *receiver* is constrained.

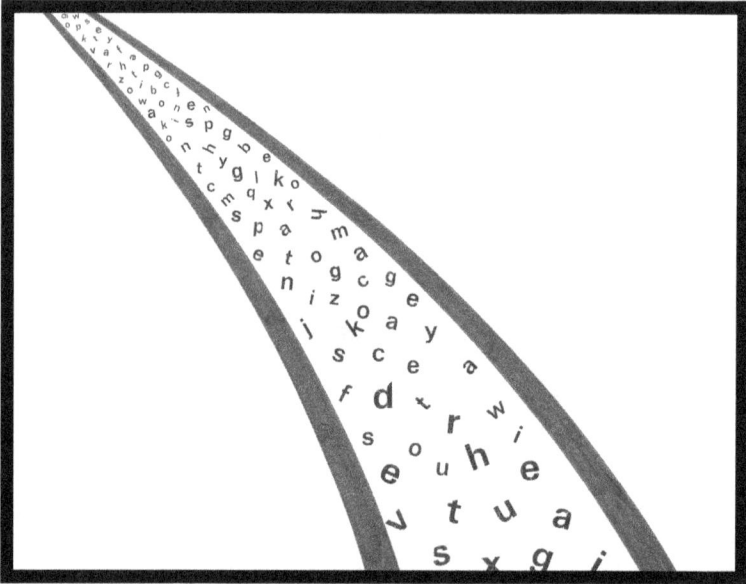

NEWS - AN UNENDING STREAM OF CONFUSION

Today, we face a world of intellectual apathy unwilling to think, to argue, and to debate issues which require reasoning. Television enhances this state and therefore is contrary to the interests of survival of the human race. *Homo sapiens* represent, at present, the ultimate evolution of the brain. To deny this evolution is suicide. To create an instrument that furthers such an attitude is both stupid and morally wrong. While we have great arguments and laws protecting our society from the debilitating influence of certain drugs, we have never posed the question of what television will do to a human race already committed to intellectual leisure. The answer is quite clear—it is the greatest brainwasher, far more than any political ideology can ever hope to be.

37

INFLATION:
NOT A VIABLE ECONOMIC CONCEPT

Since economics is a field that concerns all of us, it seems only right to make a few points. This is further justified by the fact that the experts in the field are obviously confused on the issues before us.

Point Number 1: The term inflation in itself is a misnomer. The expression would lead one to believe that there is some positive growth involved. We all know that this is not true. Inflation is, in fact, negative growth—the decline in value of money. As a result, *deflation* seems a far more appropriate term.

Point Number 2: Economists, or at least some of them, claim that in a free enterprise system the choice is between inflation and unemployment. The fact that many countries have had both for quite a while clearly demonstrates the fallacy of this argument. In fact, it seems that neither impedes the progress of the other.

Point Number 3: Inflation, and in particular galloping inflation, represents a case of unchecked growth, even though it is negative. We have previously noted that such growth can only result in collapse (see Chapter 14). The collapse will be

171

that much more devastating when the growth has been artificially prolonged, as is the case today. Governments employ methods to this end that would put any private citizen behind bars. Be that as it may, one thing is clear, inflation as a continuous process is doomed—even fourth-grade arithmetic would suggest that much. History provides the example, but as usual to no avail.

Point Number 4: Inflation further undermines the already weak moral fibre of nations. In the case of inflation, he who consumes the most and produces the least—i.e. the one with the biggest pile of debts—is furthest ahead. This may not be serious, since we have already reduced the upright or honest citizen to a fool in most western societies. But it still points to the fact that above all, inflation is ethically unacceptable.

The often heard declaration today that something is not economic is no more than an expression of laziness, rampant in the affluent societies. One must call this, simply, relative economics. There is, however, an absolute economic limit. Generally, to produce a needed raw material, energy and other raw materials are used. When this use equals the gain, we have found an absolute limit of profitability. As an example, if it takes one barrel of oil to produce a barrel of oil, the venture obviously becomes self-defeating.

Governments have no real interest in combating inflation, since they are the only real winners. What was a wealthy person's salary 10 years ago, and accordingly taxed progressively, is today an ordinary remuneration, yet it is still taxed at a wealthy person's rate. Thus, government income shows not only an inflationary, but a very real, increase due to the existing progression that is never revised. However, even for the government, the advantage is short-lived. Inflation being a form of unchecked growth inevitably results in a curve such as Curve 1 in Figure 14.2—collapse.

38

THE CONCEPT OF GOOD AND BAD

The idea of good and bad is certainly as old as man and possibly older. It is not difficult to write a doctoral thesis in philosophy on the subject, and it would be sheer madness to expect anything approaching a definitive answer in this short chapter. However, the concept is so fundamental, and of such basic importance to us, that we all must and do reflect on this topic from time to time.

In today's world, the actual definition of good is very simple: *good* represents those actions that ensure the best possible life for the individual. This is not to say that this definition is right, nor that it should be upheld. It is simply a fact of life.

One can also say that in an absolute sense, good and bad must be defined as actions that are either beneficial or detrimental to the survival of mankind. Simpson talks of the life ethic simply becoming a survival ethic. In plain terms, good is what enhances the survival and further evolution of the species, while bad is what endangers these matters.[1] In between these definitions of good and bad, centred, respectively, on the individual and mankind as a whole, are those definitions that apply to families, nations, or races.

It is quite clear that the current definitions of good and bad, restricted to the individual and the group, are on a collision course with the absolute definition in regards to mankind. It is presumably a question that can never be properly resolved, and the best one can hope for is a reasonable compromise. It is also evident that at present, no one seems to have achieved such a compromise. In the West, personal freedom is rampant, to the extent where it clearly spells doom for at least whole nations, if not mankind as such. In the East, the individual is completely suppressed, and good is defined in terms of the state (group).[2] The eastern good is just as detrimental to nations, as well as mankind, and since it also deprives the individual of all that he might consider good for himself, it is clearly the larger of two evils. The one exception may be China (see Chapter 25).

On the question of instincts, one may ask why many authors seem to feel that the acknowledgment of instincts in man is inherently bad and degrading. It is, however, evident that most of the instincts to which we readily admit in the animal world work in a far more beneficial way for these animal species than man's consciousness does for him.

39

OUR GREATEST ENEMY: OUR EGO

That our ego is our greatest enemy is neither a new, nor a profound, discovery. It should, however, not detract from the basic truth of this statement and the fact that in this regard, no progress has been made throughout man's history. It is a central question, particularly in today's crowded world, where we deal with several million frustrated egos.

Ardrey has stated that identity, stimulation, and security are the three driving forces of man and in that order.[1] Observing the world around us, it is difficult to disagree with this statement, although it may possibly be incomplete. The question of identity is, naturally, that of our ego—our desire to be recognized, to be in the limelight at least once in a while. It is for this reason that every western daily carries pages of personal pictures of salesman of the month, or of the year, and of appointments in certain companies. Usually, nobody cares except for the one depicted and his immediate family. For the more sophisticated academic and managerial egos, one creates such publications as *Who's Who*. Again, nobody cares except those mentioned, and the sales of the volumes to them assures a handsome profit for the publisher.

One can try to satisfy one's ego by establishing a truly

positive record, by outstanding work in an area, be that as an artist, scientist, or in the social field. However, the most serious workers are seldom recognized unless they are also not hesitant to advertise their own merits. This is what opens the way to our phony relationships and sets of values. Many of those whom we have met in our history books are undoubtedly unworthy of the honour, while some of those who did the most for mankind remain unknown forever. This is an inevitable conclusion as one grows older. Another, and possibly more serious, aspect of one's ego is the fact that it can also be satisfied by becoming notorious. One only needs to open the daily paper to appreciate how many try this route today.

Considering the human ego, and the fact that it is generally inflated beyond reason, also offers other possibilities. It is equally a fact that the size of the ego is almost proportional to the amount of training a person has received. Thus, academics (20 years of schooling and more) are well-known for their conceit. Since we realize that emphasizing the human ego is undesirable, we come to the conclusion that our great systems of education are really not doing the job. They totally fail to provide man with a realistic concept of his significance and thereby do more harm than good.

It is also evident that total abolishment of the human ego would mean the destruction of man as we know him. A total deflation of the human ego would in all probability mean the extinction of mankind. The task at hand is, therefore, the one shown in Figure 39.1. To determine the permissible size of the human ego and how to achieve the necessary reduction would seem to me to be one of the foremost tasks for our humanists—it is a prerequisite for the human revolution (see Chapter 47). It is intimately related to our modern predicament and unless solved, human survival is in jeopardy.

Post Scriptum
Some may argue that we always have had small groups that have successfully overcome their egos. Usually, they were posthumously promoted to sainthood. However, to claim that

MAN'S CHOICE

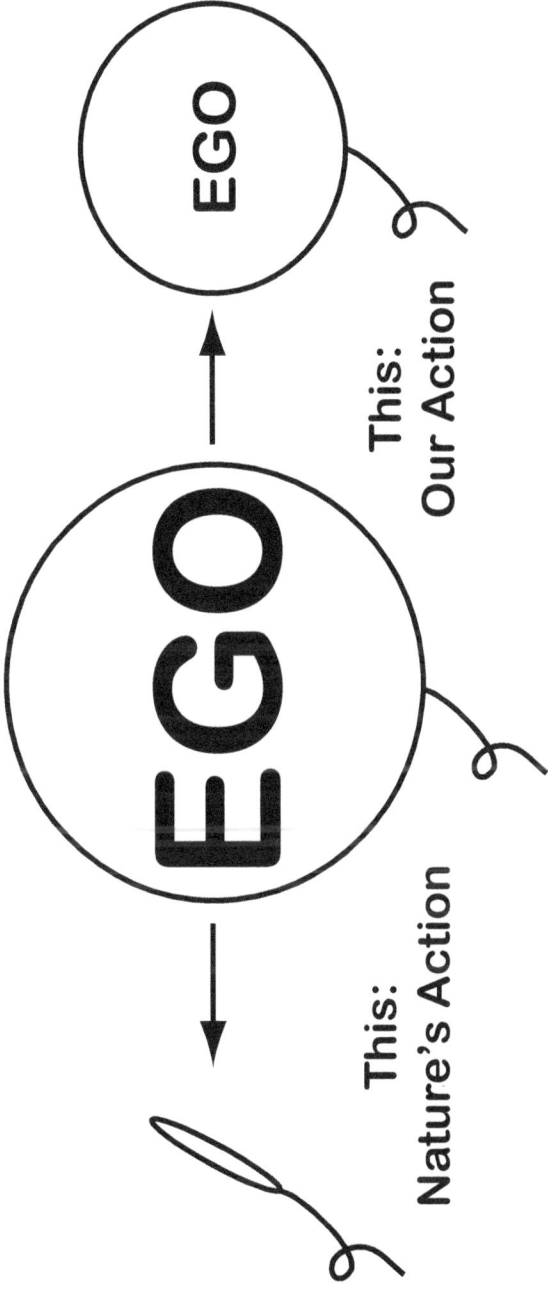

FIGURE 39.1: MAN'S CHOICE RE HIS EGO. The choice is simple: (1) take matters into our own hands and revert from an overinflated ego to a more reasonable, sustainable one; or (2) let nature dictate the outcome - extinction leaves little room for ego.

these people were entirely philanthropic, with no thought of their own wellbeing, is only partly correct. In many cases, such people devoted to the service of God are simply those willing to accept deferred payment. A reversal of the concept "fly now pay later" to "pay now fly later." Unfortunately, many of the most selfless deeds of human history when viewed in this light lose much of their glory and become far more human.

TOMORROW

40

QUO VADIS HOMO SAPIENS?
FUTURE: AN ATTEMPT AT A DEFINITION

For too long, we scientists have welcomed the past and left the future to the science fiction writers. The reason is simple. The past is safe, it is known and can be analyzed with a fair degree of certainty, while the future is unpredictable, and as Simpson says, we can only explore its possibilities.[1] This is a far more vague and less satisfactory endeavour, and one from which a true scientist should refrain. At least that is what we thought. However, man must live with the consequences of his actions, and his ability to foresee the consequences of his actions makes ethical judgment of them both possible and necessary.[2] As we have seen, man's technological development has progressed at an astounding pace in the context of earth history and yet, today, the future of *Homo sapiens* is more uncertain than ever (See Figure 40.1).

It remains for us to define the term *future*. Just what do we mean—days, years, decades or hundreds, thousands, or even millions of years? Evidently, man's concern for the future decreases in some fashion with time. We tend to be more concerned about next year than 2020 or 2030. The years 2200 or 3000 for most of us become lost in the *haze of the distance*.

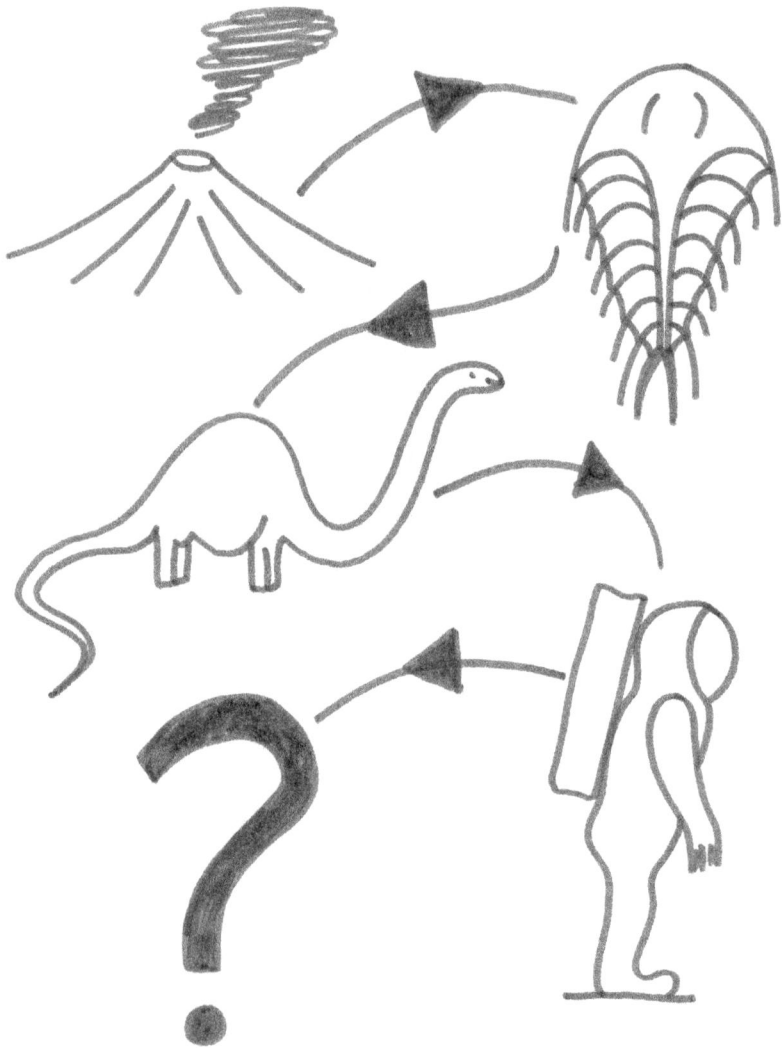

FIGURE 40.1: QUO VADIS HOMO SAPIENS?
Our children may sketch out man's evolution as above.
Amazing progress by the toolmakers over a short period of
time (geologically speaking), but the future remains
uncertain and precarious should we stay on the present path.

Such future spans exceed the imaginary capabilities of all but the most gifted. However, indications are that our present problems will not let us rest that long. They are going to be solved *by* us or *for* us within the next century.

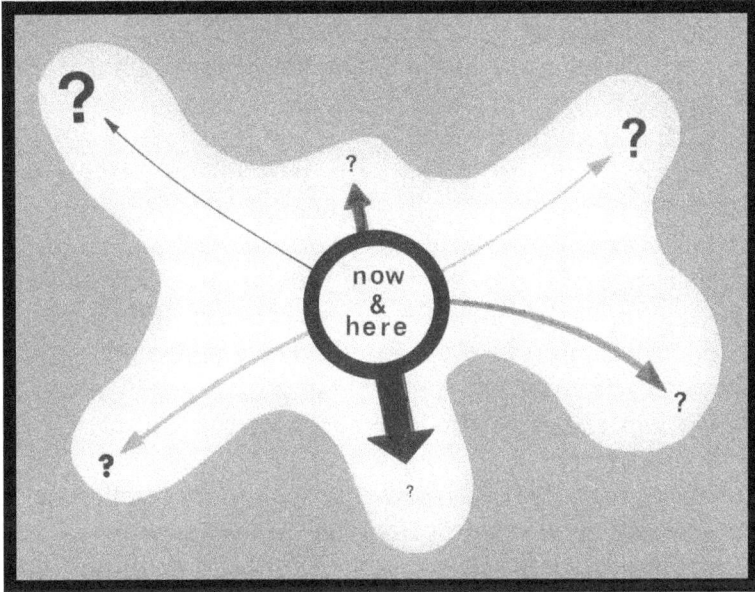

WHERE TO FROM HERE? - UNCERTAINTY

Clearly, at the lower end of the scale we cannot really talk about a responsibility towards the future. To look only days ahead is doing no better than a squirrel, which at least recognizes the seasons. On the other hand, to look thousands or more years into the future is a task to which the human mind, advanced as it may be, is simply not up to. A thousand years represents about 40 human generations. It staggers the mind to envisage a personal ancestor 40 generations ago, and it is even more forbidding to think of our successors 40 generations hence. Thus, we are left with a "reasonable" time span, which is measured in decades. And since it is this very near future that is currently threatened, or so many of us believe, it only makes sense to explore its possibilities.

One may approach the problem from a slightly different

point of view and ask what one could expect to be the concern for the future by a "normal" human being. Certainly, this is linked to the fact that the near future will be formed and lived in by those who we know—our children and grandchildren. To have no concern for their well-being would reduce us to the stage of lower animal life.

It is with this in mind that man's range of concern for the future, shown graphically in Figure 40.2, was drawn up. For those with a high level of concern, the near future—months, years—is characterized by a sharp drop-off, followed by a plateau and then, 60 to 100 years hence, another sharp drop-off to a level of concern near zero. The latter coincides with the disappearance of those whom we have known ourselves. The sad fact remains that large segments of mankind—both in the underdeveloped and in the affluent areas—have little concern for the future. One sleeps better that way, but it is an attitude that does not enhance the prospects for the continued existence of the species.

We must realize that this range of concern applies to western man only, who does not believe in reincarnation. People afflicted by a belief in a return to a future life should have a much keener and more extended concern for the future for two reasons. First, they themselves—not just their offspring—will have to live with the situation they have created. And second, since they reappear repeatedly on this earth, they will have to look further ahead than just the next few decades. The belief in reincarnation does place quite a burden on one's conscience.

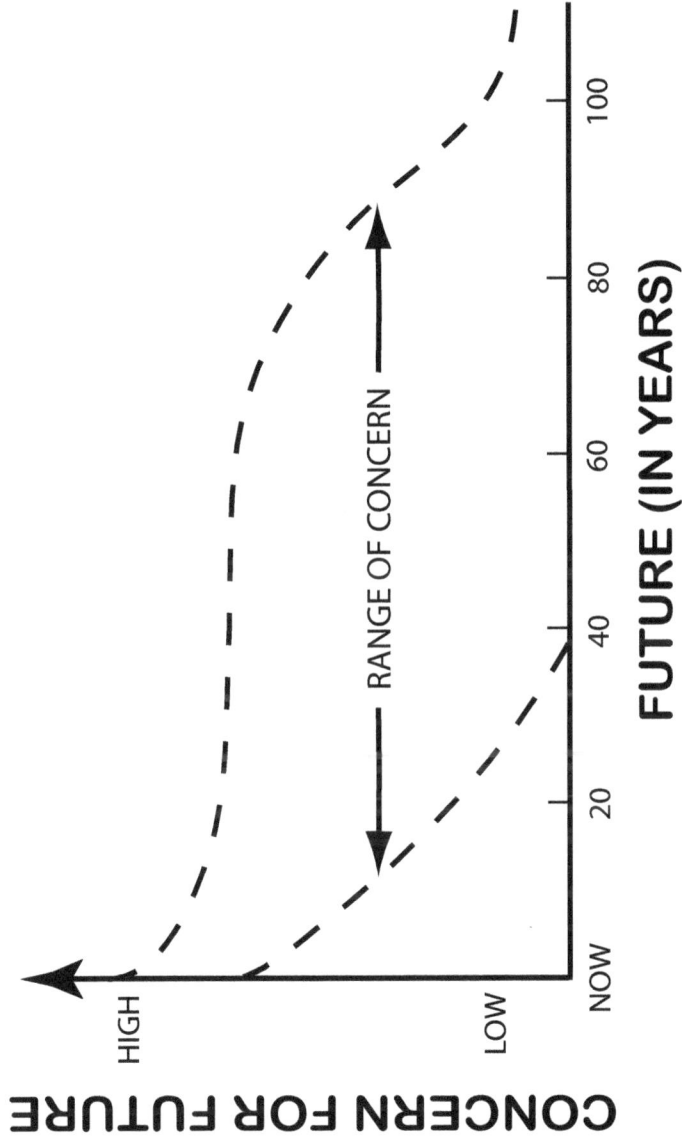

FIGURE 40.2: MAN'S CONCERN FOR THE FUTURE. An attempt to show graphically man's interest in the future. The broad band tries to do justice to individual variations.

41

THE CONCEPT OF THE ECOLOGY RISK

Today, the positions between the "developers" and the "conservationists" (or "environmentalists") are highly polarized. Both carry their viewpoints to the extreme, thereby damaging their credibility. To any outsider, it is clear that a balance must be struck. Often conservationists, in their extreme attitude, refuse to acknowledge that the present number of people simply demands development and that denial of such development in fact denies the right of the present generation to live decently, or to even live at all. On the other side, developers feel that unrestricted exploitation is not only necessary but constitutes a basic human right. Both positions are indefensible and are bound to lead to short or long-term disaster.

The current population level is so high that even if waste making is restricted to an acceptable level, certain developments have to take place. The energy needs (not demands, see Chapter 15) are such that large power plants are necessary. Large plants of any form include risks, and these risks must be accepted—they cannot be eliminated, only minimized. Thus, the concept of the ecology risk is a very real one that cannot be avoided. The stern conservationist maintains that no such risk is acceptable, while the optimistic developer refuses to even acknowledge its existence. Maybe in his heart the latter knows that he is wrong, but has a philosophy: *après nous le déluge*, i.e. little or no concern for the future (see Figure 40.2).

How to strike the balance? What constitutes acceptable risk? These are the real questions confronting us. Technically speaking, we must look for ways and means of minimizing these risks—to eliminate them is obviously a doomed endeavour. Minimizing risk also requires minimizing the demands and cutting the expectations to a minimum acceptable level. Thus, the ecology risk is a complex technical and social question.

42

TO STRIKE A BALANCE BETWEEN DOOMSDAY PESSIMISM AND IRRESPONSIBLE OPTIMISM

Reading through the current literature on modern man's predicament, the question of how to achieve a balance between doomsday pessimism and irresponsible optimism emerges as a central theme. It seems to be in man's nature to produce polarized viewpoints. We have seen this in the debate on man as a product of his nature and his nurture and we see it between the extreme conservationists and the uninhibited developers.

The call for restraint in further growth by Meadows et al.[1] figures prominently in what has been labelled doomsday literature by many. As I see it, unjustifiably so (see Chapter 14). The works of Maddox[2] and Beckerman,[3] on the other hand, are examples of what I would label irresponsible optimism. To maintain the view that the unrestrained growth of today can be perpetuated for any length of time is unjustifiable. True, as Beckerman points out, such growth has taken place in the past, but extinctions are also a fact of the past. In other words, growth will be checked, and the checking by nature involves unpleasant consequences for those being checked. It would be too much to expect any lower form of life to take preventive action. This kind of forethought requires the level of consciousness developed only in humans.

It is the opinion of those advocating a slowdown that this can be achieved without depriving those that still need growth in order to move away from the existence level. It is the affluent societies one has in mind in advocating a slowdown. It is customary today in affluent societies to subscribe to a waste of material goods which is absolutely unnecessary. Mental and spiritual freedom, the most important aspects of a dignified life, can be maintained without the presently associated high level of material consumption. On the contrary, the material consumption, and the energies directed in the production and consumption of these material goods, detracts from enjoying the more leisurely aspects of life.

Thus, curtailment of growth is envisaged as taking place from the top. To drop the A_{max} to 20 and bring the A_{min} to say 10.* This might result in an average A of 15 and with the present population of 4×10^9, we get an effective population (EP – see Chapter 15):

$$EP = 15 \times 4 \times 10^9 = 60 \times 10^9$$

while our present situation (see Chapter 15, Equation 4) is as follows:

$$EP_{now} = 1 \times 3 \times 10^9 + 30 \times 1 \times 10^9 = 33 \times 10^9$$

Thus, we have only a doubling of the effective population under these conditions, rather than a four-fold increase as in our previous computation in Chapter 15, where we calcu-

---

* To strive for a uniform A=15 would be both unrealistic and socially unjust. Different types of work must be rated differently. Every individual has a choice: an easy, risk-free job with little demand and therefore little reward or a demanding, risky job with concomitantly higher material rewards.

lated the impact of bringing everyone to the present American standard of living.

None of this alleviates problems altogether, but it buys time, which is crucial. Changes, whether they be technical or social, do require time, and we must be clear that any interruption in supplying the needs of the population has devastating consequences. It is of little consolation in 2010 to know that unlimited energy will be available in 2020 when the period in between is not covered.

One other aspect one must not forget is the cumulative effect. A modest conservation effort by an individual may only inconvenience him or her. Also, his or her savings may seem of insignificant proportion. However, the savings must be multiplied by a large figure, such as 10^9, and therefore that does become significant on a worldwide scale.

Those that cling to the present definition of the "good life," and believe that it can be maintained in unchanged form, are not realistic. On the other hand, the pessimists are not necessarily doomsday sayers. They advocate changes that, once initiated, may have many beneficial effects and that will enhance the prospects of the western societies surviving the next few critical decades. The continuation of present trends spells doom for the western societies. It is these societies that today have the option of disappearing quietly, as others have in the past, or with a bang, endangering the whole of mankind. The possibility that they choose the bang cannot be ruled out.

43

THE 21ˢᵀ CENTURY: A GEOLOGIST'S PERSPECTIVE

Having determined in Chapter 40 that a reasonable horizon of human concern is in the range of decades, we shall stretch this a bit here and have a look at the 21ˢᵗ century. Geologists are as mesmerized by the 3ʳᵈ millennium as the general public. Their forecasts—as one would expect—are focused on petroleum. The following discussion provides a more general viewpoint, with emphasis on geology.

Homo sapiens live off this earth. Dollars, Marks, Yen, and Swiss Francs are only viable as long as they remain convertible to "something useful." Potable water, food, and energy are the primary requirements for our existence. All are provided by our planet or, rather, accumulated in its outermost skin. Geology, therefore, is *the* basic science in our lives.

The dawn of the newest century—artificial as it may be—marks a turning point insofar as we leave the *century of the locusts* to enter the *century of the crunches*. The former is a time of efficient or ruthless (your opinion) resource exploitation. The question poses itself: "What really drives the demand for resources in the developed countries?" As we have seen, according to Ardrey the parameters that shape the behaviour of *Homo sapiens* are identity, stimulation and security.[1] Considering those around you in these terms saves acquiring a PhD in psychology.

Clearly, stimulation and security are often on a collision course, one being at a maximum when the other is at a minimum. Stimulation is the opposite of boredom, with the latter all too often afflicting our highly intelligent species. In fighting boredom, we blow away more resources between Friday night and Monday morning than during the entire working week. Conclusion: the enormous resource appetite (and concomitant pollution) of the overdeveloped countries is largely fuelled by our desire for entertainment and is not required for a basic, comfortable existence.

In the century of the locusts, world population has risen from about 1.5 billion to 6 billion people. However, this is *not* the whole story. As we saw in Chapter 15, contrary to the animal world, the impact of the human race cannot be measured by numbers alone. Today, various populations make very different demands on the earth's resources. Assume that: (i) the average demand in 1900 was measured by an *affluence index* of 1; (ii) the current affluent West (roughly 1 billion) lives at an index of 10; and (iii) the remaining 5 billion today still live at the minimal index of 1. This gives us an *effective population* of 15 billion. Now assume that we want to elevate the whole world to the affluent western index level. That would give us an effective population of 60 billion. Compare that to the 1.5 billion in 1900 and one does not have to be a geologist to understand the impossibility of this scenario. *The Limits to Growth* (see Chapter 14) may be flawed in detail (specific computer models) but cannot be argued in principle. Remember, the above computation *is* conservative.

The coming crunches are not restricted to oil/gas, even though we live in what aptly might be called the oil-age after passing through the stone-, bronze-, and iron-ages. *Water, oil/gas,* and *social tensions* will all be subject to crunches, which will become critical somewhere near the middle of the 21st century.

Water may well precede oil/gas. Fresh water exists only as a thin skin on the continents in the form of lakes, rivers, glaciers, and groundwater. The total reservoir available is limited and has been reduced in the past 100 years by overuse and contamination (see Chapter 44).

In terms of *oil/gas*, the *Hubbert curve* (peaking oil production represented by what is essentially a parabolic curve) is still valid.[2] No doubt, King Hubbert will be recognized by historians as one of the important contributors to geology in the 20th century. He made it quite clear that the area under his curve is open for discussion. In fact, he himself operated with both a conservative and an optimistic forecast. However, the basic nature of his curve, valid for a limited resource, is not debatable. The *Hubbert curve* will be modified in that it will be skewed with a long tail. Furthermore, it is important to realize that the crunch does not coincide with the tail-end but rather the turnaround when demand outstrips production. Hubbert's optimistic version of 2,100 billion barrels of recoverable oil is still viable and represents an amazing achievement (not all doomsday sayers are quacks).[3] The arrival of this crunch can also be placed near the middle of the 21st century.

It is interesting to contemplate that for the U.S. to replace its 20-million-barrel-per-day oil "fix" would require about 1,500 major power plants, each with a 1,000 megawatt installed capacity, either nuclear or coal-fired. Since this is electricity, it would also necessitate a complete rebuilding of the transmission infrastructure.

Social tensions are already rampant in our crowded world (for details, just consult your favourite daily). By the middle of the 21st century, the social fabric of most countries will be reduced to shreds. Man, with his superior consciousness, is a lousy social animal to begin with. Under crowded conditions, he becomes self-destructive.

To geologists, these observations and concomitant warnings do not come as a surprise. Life on earth has always been risky and subject to change, often at a frantic pace. The climate, as well, never remains the same. The alternative to global warming is global cooling, accompanied by the rise or fall of sea levels and the build-up or waning of ice-caps.

Provided any trend persists over any length of time, there is only one question: "Who acquires refugee status, the Dutch or the Canadians?" Neither is an attractive prospect for an overcrowded world.

Green movements are hell bent on saving the planet. Well, let's just be realistic. The *planet* is doing just fine. True, it has a little skin cancer, deadly to the human race, but hardly of any consequence to the planet.

To sum it all up: it is not just oil/gas, or for that matter shareholder value, or money that counts. In the basic analysis, our existence is based on geology. It determines where the resources are and how they can be developed efficiently and responsibly. It also dictates the distribution of renewable harvests of any kind, such as: water, wood, grains, fish, etc. Yes, geology, after all, is the *mother of all sciences*.

Editor's Note
While the balance of this book, save for Chapter 45, was written in the late 1970s, this chapter was penned in February of 1998, as the world was preparing for "Y2K" and the promise of a new century.

44

THE INTERDISCIPLINARY DIALOGUE: THE CHALLENGE OF OUR TIMES

So how do we move forward? What can be done to address the challenges we have reviewed over the last 43 chapters? Simply put, we need to stop the denial and start engaging the real issues facing us. And we must address them in a global manner. An elegant fix to part of the problem may well prove futile if other parts are ignored (recall the analogy of the multi-part truth – Chapter 11). The enemy in achieving a global solution is the fragmentation of knowledge (see Chapter 12).

The interdisciplinary dialogue[1] is a step towards addressing this fragmentation and has two most important aspects:

1. Without it, a university is a mere trade school, an institution of higher training but not a *Universitas* where young people are encouraged to make the individual commitment to qualify for the status of being "educated."

2. It is vital in terms of modern man's predicament. Unless such a dialogue can be initiated and brought to a very intense level, man will turn out to be an

evolutionary failure. There is, of course, nothing unique about the latter, as any student of ancient life knows.

Definition of the term "Interdisciplinary"
Before we embark on a discussion of the subject proper, it may be to our advantage to clearly define the type of dialogue we are seeking. Today, we have reached such a degree of specialization that we are not only in need of inter but also intra disciplinary communication. I would submit that even at that latter level, much remains to be done. However, this certainly does not concern us here.

The interdisciplinary dialogue itself can take place at various levels. No discipline is fully self-sufficient. This is particularly true of geology, my own field of study, where we lean heavily on all other natural sciences such as physics, chemistry, biology, and as a basis for all, mathematics. Naturally, there are also strong ties to various branches of engineering. After all, exploration for and exploitation of natural resources is the main practical aspect of geology. On the other hand, it is not difficult to see that certain fields should have strong ties to geology on purely academic grounds. Look at economics as an example—the availability of natural resources is fundamental to modern man's life. And who, but the geologist, compiles the inventory!

Another case is presented by history. I ask you: "Does it make sense to explore the short span of modern man's history without ever measuring it against the background of the history of our planet, or at least the history of life as we know it today?" I submit that such an approach might induce historians to take a somewhat humbler view of our species, much to the advantage of us all.

Leaving my own field out of the picture for the moment, one might postulate strong ties between anthropology and philosophy, two specialties at, or very near, the heart of any university. One cannot ask basic questions about man and his destiny without full knowledge of his early—now measured in millions of years—history and his animal heritage. In regard to the latter, it is not only the anthropologist that becomes involved, but it is most fascinating to follow the accounts of modern biologists. After tearing themselves loose from the exciting task of labelling bones, biologists have, amongst other things, begun to study animals in their natural habitat. The results are most astonishing to a form of life that has habitually considered itself much divorced from the rest of life and, either openly or deep down, has never totally rejected the possibility of divine creation.

How close are our ties to the animal world? Does man have instincts? These are questions that are not presently resolved, that are vital to an understanding of ourselves, and that concern all of us. Unfortunately, these topics cannot be approached with the same detachment as rocks or the law of physics. Where man himself is involved, and we look essentially at ourselves, we are apt to become emotional and our unbiased scientific minds tend to be shrouded in the smoke that rises from our burnt egos. A point to keep well in mind, when as a natural scientist one tries to talk to a social scientist or humanist.

Again, all these are examples of a limited interdisciplinary dialogue, without which it is really impossible to function properly in one's own discipline. The dialogue to be discussed here is the one on the highest plane and on the broadest level between those two cultures we have talked so much about (see Chapter 11):

THE HUMANIST and THE TECHNOLOGIST

With both terms to be taken in the broadest context.

The *humanists* are all those dealing with man, his value systems, his motivations, his actions and reactions, as well as the question of human instincts, and the effect of the cultural versus the biological evolution.

One of the central questions: "Why is man so aggressive? Where is the root to this trait to be found?" In terms of our further discussion: "In what respect do the biological and cultural evolutions differ?" One possibly trivial, but nonetheless presently vital, point, is that the rate of change in the latter can be infinitely higher than in the former.

The *technologists* are those who are essentially concerned only with one aspect of human nature: the extraordinary capability as a super toolmaker. Specifically, this group will include the natural scientists and engineers.

There can be little doubt that at present there is very limited interaction between these groups, both in the university environment and in daily life. Certainly, the humanists employ with enthusiasm many gadgets provided by the technologists. They do not, however, interfere with their invention or production. If anything, they may spur the technologists to ever higher "achievements."

The technologists have altered the consequences of certain human actions drastically by delivery of better and more powerful tools (or should we say "weapons" as Ardrey[2] suggests?). But technologists have made no effort to determine man's destiny, except in the indirect way outlined above. Worldwide, they refuse to be involved in anything except their field of specialization, an attitude that can only be termed criminal in view of the current situation. That this is not a distinctive western feature we may infer from Solzhenitsyn's writings. In his Nobel Prize address, he says:

> But no; scientists have made no clear effort to become an important, independently active force of mankind. Whole congresses at a time, they back away from the sufferings of others; it is more comfortable to stay within the bounds of science. That same spirit of Munich has spread its debilitating wings over them.[3]

One might add here that scientists have repeatedly claimed that they can contribute to an understanding between people living under different political and economic systems, since science knows no boundaries. They forget that ethics and resulting moral rules are even more important. Only by disregarding ethical concepts entirely has it been possible for scientists to get along so well internationally. Their claims must be considered null and void.

The Present Predicament
There can be no denying that, on the whole, technologists have been far more successful than those who try to understand and modify man's behaviour (see Chapter 13). One can advance a number of theories as to why this is so, but the existence of this imbalance in human development can hardly be disputed. The current mutual indifference is not likely to reduce the gap between *man the man* and *man the toolmaker*. The isolationist attitude in both camps is difficult to understand in today's highly technological world, with problems largely or exclusively of man's own making. This imbalance in development is the basic reason for modern man's predicament and Simpson has phrased it this way: "The inequity of knowledge is in itself unethical and is one of man's great blunders. It could be his last one."[4] There can be little doubt that human tool making today constitutes a classic case of *hypertely*, i.e. evolutionary overdevelopment to where an originally beneficial trait becomes detrimental to the further existence of the species.

We have alluded repeatedly to a predicament. We refer to the by now well-known aspects of our modern, affluent world such as: unhealthy increase in population; depletion of non-renewable resources; intolerable pollution; and threat of atomic annihilation.

The existence of these problems is hardly disputed, and it is encouraging to see some agreement, however small. What is ardently debated is the seriousness of these problems as well as the relative priorities. It is the battle of the optimists versus the pessimists (see Chapters 19 and 42). The former, represented by people such as Maddox,[5] see all problems as relatively minor—nothing that a little more technology cannot solve. The latter's view is well put by Engel:

> The optimists believe, or claim to believe, that ours is the best of all possible times in the best

of all possible worlds. We pessimists are those
who fear that the optimists are right, and that
man, if not the earth, is running out of time.
Here, at two thin protruding points from otherwise polarized positions, the attitudes of the
pessimistic scientists and Jehovah's Witnesses
almost touch.[6]

Once we agree that problems exist and are serious enough
to warrant action, then those with higher training have an
advantage and the responsibility to take up the battle, the
cause of which must still be hidden to many. Despite much
recent publicity, the issues are subtle and not easily recognized. In light of this, it is difficult to understand the smug
attitude that still prevails at western universities. The discussion about man's future should intensely occupy all disciplines
and students and faculty alike.

Naturally, "to be concerned" means to accept a responsibility to the future (for a discussion of what is meant by
the *future* and our collective responsibility to that future, see
Chapter 40). But is this not one of the main features distin-
guishing man from animal? Simpson says, "It is the capacity
to predict the outcome of our actions that makes us responsible for them and that therefore makes ethical judgment of
them both possible and necessary."[7]

Growth as a Main Evil of the Present Predicament
Many of the previously mentioned aspects of the contemporary
predicament of man are related to unrestrained growth in a
number of areas. Growth is a cherished concept in our society
and many others. Only recently has the validity of this idea
been challenged. It does not speak for the intellectual calibre
of western universities that it took us so long to recognize the
obvious. It is also most deplorable that now those who can
rationalize the danger are not putting themselves behind the
defeat of the concept with all the necessary vigour.

The honour of the academics has been salvaged to some
extent by the MIT group with *The Limits to Growth*, sponsored

by the Club of Rome. See Chapter 14 for a more detailed discussion of their work.

Beyond computer models and technical studies, there are simpler indicators of overpopulation. For example, around 1850, the Americans sent the Mormons into the desert. That worked. The reason: too few Native Americans. Around 1950, the world sent the Jews into the desert. That did not work. The reason: too many Palestinians.

The pressures of a growing population in a finite system are all-pervasive. Take, for example, the impact of population on global climate challenges and the fresh water supply.

History tells us that climate is not a constant. Warm and cold periods alternate. In each case habitable regions are opened and closed, which leads to mass migrations. In an overpopulated world, such migrations are difficult to accommodate. The core problem is not the climate but overpopulation.

Water, in the sense of drinking water, only appears on the continents in the form of glaciers, lakes, rivers, and groundwater. These reservoirs are constantly replenished through the hydrological cycle—that great desalination "plant" which delivers pure water to the continents in the form of rain and snow. One should never forget, however, that this replenishment has boundaries, and the reservoirs are limited. For example, potable water is restricted to a depth of approximately 500 metres. At greater depths, the groundwater is too saline. Population pressures leading to a dramatic lowering of the groundwater table in many areas, and increasing pollution of the groundwater, make it likely that access to fresh water will become a critical issue in the 21st century.

It is one of the foremost tasks of tenured academics to pound into the western business world that only doom and gloom can result from unrestrained material growth. And believe me; you need tenure to express such heresy. At this time, a word on academic tenure seems in order since it is intimately tied to the interdisciplinary dialogue, which involves a severe critique of current social systems.

Academics have consistently claimed that tenure is the most viable way of protecting academic freedom. What they have equally consistently overlooked is the fact that the concept of academic freedom is both a privilege and an obligation. The public has every right to expect an academic to be more than just a capable professional. The latter is simply a prerequisite, but not the only requirement, to qualify for what one calls a "professor." The other, equally important, aspect of the professorship is the commitment to intellectual endeavour, to inquiries about the nature of man, his societies, and his destiny. This inevitably leads down the road of becoming a critic of the present social forms (it always has). At that time, academic freedom is used and tenure becomes a defensible concept.

The academics in the western world have, by and large, failed in this regard. This is painfully obvious when considering the present predicament. It could not have arisen if the academics had taken the role in society they ought to play by virtue of their training and (hopefully) superior intellectual capabilities. Each one of us must make his own assessment of whether or not tenure for academics is, indeed, a concept that can be supported.

The authors of *The Limits to Growth* also call attention to the fact that modern problems tend to be related. As an example, industrial production leads to both the exhaustion of non-renewable resources and pollution of the environment. One may again term this a rather trivial observation. But many of those zealous "world improvers" better take note that you cannot tamper with one part of the system without prompting a chain reaction, or in other words, you must look at the whole picture.

The Cause of Growth
The problem of sustaining man on this earth is not only related to his numbers per se, but also to his style or standard of living. Chapter 15 deals in depth with this principle and the concept of the *effective population* (EP). Simply put, an affluent person in the developed world places demands on the earth that are multiples of those imposed by a poor citizen from an underdeveloped country. The clear message is

that it is the effective population that counts and not merely the total number of people present on this earth. The EP is a function of our *technical possibilities* and our *human attitudes*. The advances in technology provide the options, but it is not necessary, and often not desirable, for us to exercise these options. However, at present we do—with no questions asked.

The exponential growth curves are one typical example of modern man's predicament and they demonstrate that the current problems are related to the inequity in wisdom and technical skills in man. Unless the balance can be restored, evolutionary failure is inevitable. This is something that concerns all, and the interdisciplinary dialogue is no longer just one of the intellectual niceties a real university offers—it becomes a matter of survival.

The Human Factor

As we have noted previously, Robert Ardrey once stated that three factors motivate *Homo sapiens*: identity, stimulation, and security. In my estimation, this simple rule offers more insight when observing those around you than a PhD in psychology. And as we noted in Chapter 11, C.P. Snow states that we live alone. Put these two together, and it is not hard to conclude that the interdisciplinary dialogue is *not* a natural activity for humanity. It does not come easily to an ambitious, aggressive, and egocentric animal such as man. It cannot and never will be a widespread and universal activity. In order to bring about a minimum level of interdisciplinary dialogue, we shall have to outwit ourselves.

Interdisciplinary dialogue is an unhealthy activity for both identity and security. Bill Fisher has said, and rightly so, "In synergism, the individual contribution is not recognizable."[8] So much for identity. Security is highest within one's own domain, both geographically and intellectually. To expose oneself to different fields of expertise is tantamount to skating on thin ice; again, a hazardous activity.

Stimulation certainly occurs and is, in fact, the main reason for such a dialogue. This type of stimulation is an intellectual challenge calling for much homework to be done, and it is not to be confused with the stimulation provided by the entertainment industry. The popularity of the TV serials is unique in that they manage to generate high stimulation without sacrificing security. The immortality of the lead characters assures protection, even in the most desperate moments. Stimulation from the interdisciplinary dialogue leaves us fully vulnerable, while trespassing on unfamiliar ground. In addition, it requires our active participation and that in a society where exercising all kinds of muscles is popular, but any activation of the brain is frowned upon. Participation in the interdisciplinary dialogue is both strenuous and bound to cause a certain amount of anxiety, with everyone being "out of their depth" from time to time. In short, both identity and security are not well served by this activity.

All technical and terminology problems pale before the very serious obstacles presented by our nature. It is for this reason that any grandiose plans for happy and harmonious interdisciplinary communication on a large scale are doomed. Synergism is no doubt a desirable goal. To achieve it, a series of small steps, each with limited effect, appears to be the way to proceed. To envisage a full and permanent cooperative effort is unrealistic and unachievable. My final recommendations rest on this premise.

The initiation of the interdisciplinary dialogue faces problems similar to those in establishing a good orchestra. You need top players; yet, are such individuals satisfied to be but one violin in a hundred member orchestra? In fact, the task is more formidable in that the interdisciplinary dialogue team contains only the principals from each instrumental group.

The Requirements for the Interdisciplinary Dialogue

We may reiterate once more: *The problems confronting us today concern all.* As Dürrenmatt says, "What concerns everyone can only be resolved by everyone" or its corollary, "Each attempt of an individual to resolve for himself what is the concern of everyone is doomed to fail."[9] However, those with superior schooling must take the lead. Modern man's predicament is not as obvious as being at war, even though the consequences are far more serious. The present situation is not easily analyzed, and even after concentrated mental effort, it is not easy to identify the most urgent priorities with any degree of certainty.

Our times must be most confusing for the young and those with little schooling. They are constantly bombarded with contradictory statements from people with equal credentials—be that a doctorate, a professorship, or a high position in industry, government, or politics. We need to discuss matters, set our house in order, determine the priorities, and present a united front. The present haphazard approach results only in much talk and total inaction, a luxury which we can hardly afford.

The first, foremost, and possibly only requirement is *time*. Academics must revise their priorities. Time must be taken from research projects (often more pedestrian than one is willing to admit) and reassigned to the exchange of ideas with colleagues in other disciplines. It also requires goodwill. It is not easy for humanists and technologists to converse—just consider their different communication styles, as witnessed in typical faculty lectures.

Generally, modern universities place too much emphasis on the external activities of their teaching staff. To publish, to travel as far as possible, and to generally "put the place on the map" are given high priority. Such an environment is not conducive to the interdisciplinary dialogue. To produce a more favourable climate is not really a technical matter but rather one of directing human behaviour. Peace—a certain slowing of the pulse of the institution—is a necessary prerequisite. In doing less one is achieving more. One has to take time to sit down together, accept each other as equals, and forget about academic jealousies, which are all too prevalent. In the old days, some of the best advances resulted from informal conversations held over beer or coffee in a smoky "hole in the wall" near campus.

Possible Topics for the Interdisciplinary Dialogue

1. Is there a saturation level of affluence (A_{max} – see

Chapter 15)? If so, can it be defined, what is it, are
we above it?

2. The quality of life versus the standard of living (see
    Chapter 27). What is the relationship? Can the
    two be clearly defined or are both entirely individ-
    ual matters? Is *our* quality of life what we think it
    is? Figure 27.1 represents an attempt (no more) to
    depict the relationship between the quality of life and
    the standard of living. It indicates that unlimited
    material wealth does not provide a high quality of
    life. Whether or not this curve (representing a per-
    sonal opinion) has any merit is really not import-
    ant, but we must come to grips with this problem.
    In order to understand it, we must also look at how
    affluence is acquired. Figure 15.4 shows the *road
    to affluence*. While there can be no doubt that at
    present only a minority of mankind can be termed
    affluent, the question may be raised whether there
    exists any political or economic system that is not
    somewhere on the road to affluence with accelerat-
    ing material growth. One also recognizes that over-
    developed nations will have to negotiate a U-turn on
    the dead-end part of the road, a most difficult man-
    oeuvre. Note also that while underdeveloped groups
    only represent a danger to themselves, overdeveloped
    nations are a menace to the whole species.

3. Is it possible to *screen* technical advances? The
    necessity to make ethical decisions in the field of
    technology can no longer be ignored. Skinner states:
    "Decisions about the uses of science seem to demand
    the kind of wisdom which, for some curious reason,
    scientists are denied. If they are to make value judg-
    ments at all, it is only with the wisdom they share
    with people in general."[10] True, but this is correct for
    all academics. If we are to ask about technological
    innovations not only: "Is it feasible?" but also: "Is

it advisable?" then human wisdom must be raised above its present level. It is the task of enhancing personal responsibility. As Simpson says, "In the last analysis, personal responsibility is non-delegable."[11] No system can be any better than the individuals that devise and sustain it.

4. Are there any viable political and economic systems in existence at present? Can any system be invented that is more likely to assure the future of humanity than those presently in operation?

5. The concept of the *ecology risk* (see Chapter 41). With the present high population level and corresponding needs, such risk must basically be accepted. Overzealous environmentalists have done more harm than good. To deprive the existing generation of its needs is simply unrealistic. One must try to anticipate failures (improbable events are not impossible) and minimize possible consequences. Total elimination is in many instances not feasible and only discredits a reasonable compromise. Technical know-how and human cooperation are basic requirements. The latter is currently low. Discipline runs against the grain in a permissive society, concerned primarily with individual rights rather than obligations. A permissive society in an overcrowded world is an asinine concept. It is to be expected that social and natural scientists will be sharply divided over such issues, if the latter ever decide to make themselves heard.

6. Are any of man's present ethical concepts valid and applicable in today's technical world? In particular, we might ask whether the Christian ethic (which is not currently in operation) is a workable guideline for a responsible life today. White has certain reservations in this regard:

> Both our present science and our present technology are so tinctured with orthodox Christian arrogance toward nature that no solution of our ecological crisis can be expected from them alone. Since the roots of our trouble are so largely religious, the remedy must also be essentially religious, whether we call it that or not. We must rethink and refeel our nature and destiny.[12]

7. Does history repeat itself? What can be learned from the past in regard to the future? Hegel states: "What experience and history teaches is this—that peoples and governments have never learned anything from history, or acted on principles deduced from it."[13]

Recommendations - Academia

In order to promote the interdisciplinary dialogue at universities, the top management will have to change its strategy. Currently, extra-university activities are highly rewarded while intra-university efforts are taken for granted. For example, a talk given at a departmental seminar—the fragile lifeline of intradepartmental cohesion—hardly rates a mention in the annual faculty report. The *same* talk given in Paris, or better yet Beijing, produces a rain, if not deluge, of brownie points. As long as professors are rated by the k-p-d system (kilometres travelled, pages published, and dollars raised), interdisciplinary communication is a non-paying proposition.

In my view, the reputations of universities rest on the quality of their graduates rather than the research done. Professors, through their teaching at the cutting edge, are predestined to be current in their subject areas and to synthesize the connections that become obvious in the course of teaching the subject matter. A limited amount of research is the natural outcome of competent university teaching. Good teaching requires reading far more than writing. The *publish or perish* syndrome is nothing short of insane. You disagree? Consider this example. King Hubbert, a forerunner and great contributor in both geology and geophysics (see Chapter 43), lived to the ripe old age of 86. Thus, his life included an active career as a scientist of 50 to 60 years. During that time he published numerous papers. However, the milestone papers, published in two separate volumes, number about 6 to 12. It seems to me there is a message here.

The tasks of efficient resource development and environmental protection require the formation of interdisciplinary teams. At the student level it is important to impress on people that the days of our unreserved captivation with the specialist are over. There is no attempt to exterminate the specialist, but rather the goal is to restore a *balance*. Specialization, without a general concept of the overall picture, yields only limited and disjointed results. All specialists must spend a small amount of time familiarizing themselves with the essence of neighbouring fields. See Chapter 29 for a proposal aimed at producing a true university degree (i.e. one that incorporates a broader exposure to the world as a whole). The short courses of various continuing education organizations will no doubt also play an important role in this regard.

In addition, it must not be forgotten that even the *intra*-departmental dialogue leaves much to be desired in most schools. Professors that habitually stay away from such seminars—and there are those—should bear the consequences imposed by the administration. Even academic freedom has its limits.

Recommendations - Industry

The discussion of the human factor has made it clear that the task of putting together the interdisciplinary team is not an easy one. Before evaluating the pros and cons, it is necessary to address the currently prevailing obsession with *excellence*. Excellence by definition cannot be all pervasive. It also cannot be bought. The availability of large funds does not in any way guarantee instant excellence. One must only remember that many great discoveries have been made under highly adverse conditions. Excellence *next door* is not *excellence* but *competition*. Just consider a candle burning in your office when your neighbour has a 500 watt floodlight going. Excellence only acknowledges other excellence when it is safely removed in either space or time, from which it follows that the best excellence is dead.

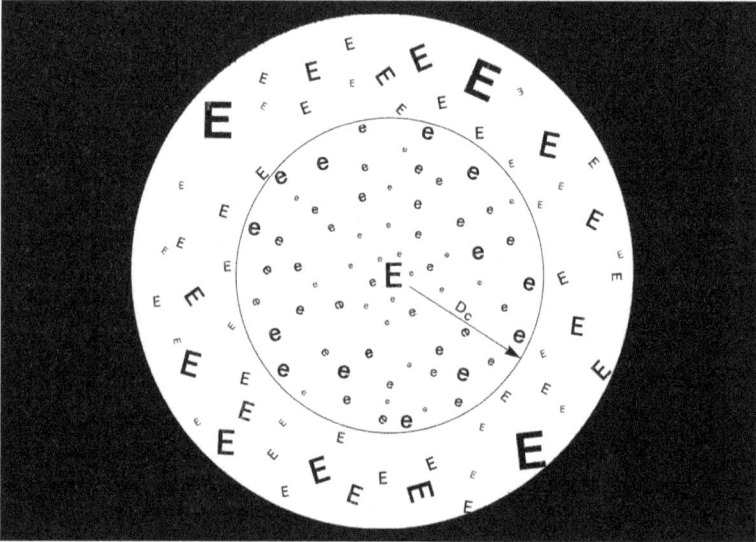

THE CENTRAL POSITION AND
THE CONCEPT OF EXCELLENCE IN SPACE
(D_c = CRITICAL DISTANCE)

One should resist the temptation to assemble a team of pure excellence. Six alpha animals around a table constitute a critical mass likely to get lost in the process of mutual haggling over the pecking order. The team must contain some "buffer material" that may be less creative but can earn its place by diffusing what become fruitless arguments and is smart enough to support the good ideas when they surface. It is absolutely mandatory for the managers of such a dialogue to follow sound management practice and to *know their people*.

One should also avoid creating standing committees and instead work with ad hoc committees. Assemble a group for a specific task and after successful completion, dissolve the group and reassemble with a different composition for the next challenge. To create "marriages" or "families" is not the way to go—temporary liaisons are better suited for the job.

Professionals operating in interdisciplinary teams will find themselves in a constant state of flux, migrating from team to team with different tasks. There is no reason why some should not hold membership in several teams at any one time. Managers will have to keep a close tab on things and separate "immiscible" excellences before much damage is done.

Excellence and egocentricity are, unfortunately, closely linked. Some excellences are constantly surrounded by crackling electricity, and a relaxed conversation—the essence of a successful interdisciplinary dialogue—is just not possible. They must be kept in a special "cage," and the manager assumes the role of zookeeper. This may sound ridiculous to some, but the fact that in most European universities the geosciences, for example, are split into so many departments—much to their detriment—speaks for itself. They learned early on that when you have two male tigers and keep them in the same cage, you soon only have one left.

Conclusion

One last reminder for both academia and industry—the basic ingredient for good science is *creativity*. It does not flourish in a pressure cooker environment, such as the one we have generated in academia and industry labs alike.

On this "optimistic" note—optimistic in the sense that it shows how much remains to be done—we shall rest our case for the necessity of the interdisciplinary dialogue.

Editor's Note

This chapter was not in the original book manuscript but was a separate paper, presented on a number of occasions. I felt it appropriate to include it in Part 3 of the book, where we take a glimpse into the future.

By advocating for the interdisciplinary dialogue as a part of "how much remains to be done," the author is signalling some optimism for the future should *Homo sapiens* be willing to take the necessary steps to get there. It is possible that a meaningful and widespread interdisciplinary dialogue will pave the way for a human revolution (see Chapter 47). We might borrow a phrase from President Obama and call this the "audacity of hope."[14]

45

STEWARDSHIP OF THE EARTH: HOW TO DEAL WITH ENVIRONMENTAL CHALLENGES IN A SENSIBLE AND RESPONSIBLE MANNER

Everywhere in the western world, *environmental science* courses, departments, and divisions are formed in order to participate in what is perceived to be the challenge of the future. The very wording demonstrates a failure to grasp the fundamental nature of the problem. It is equivalent to the movement "Save our Planet." The planet is doing just fine, and the minor skin cancer it has developed will in no way affect the future existence of this planet. It is not the planet we wish to save but rather our personal and collective existence, which is quite a different matter.

Environmental studies, and the minimizing of ecological risks, are more a matter of commitment and attitude than any new and yet to be discovered science. Environmental science—or engineering—is at best a misnomer and at worst a fraud. It implies that those engaged in it apply, for example, laws of physics unknown to the ordinary physicist. Nothing could be further from the truth. The environmental problems are ordinary technical problems to be solved by known, or to be developed, technical methods.

Science, or rather applied science (engineering), has been the driving force of the scientific revolution, the successor to the industrial revolution after World War II. Over the last few decades, science has been pushed ahead with little or no regard for the broader (environmental) consequences. As a result, the public today regards science as both the glory and the curse of our time. Even hard boiled technologists can no longer deny the emergence of serious problems.

Worldwide, the positions of environmentalists and hard line technocrats are highly polarized. This has resulted in much talk and little or no action. The latest example of this deplorable situation is the *Rio Conference* of 1992.

Currently, we are witnessing an explosive growth in environmental sciences courses of dubious scientific value and heavily diluted with social science content. This development is wrong-headed and should not be pursued.

Contrary to science, social science has been outright destructive and is largely responsible for the decline of the social fabric in all western countries. To expect social scientists to find solutions to basically scientific problems is ludicrous. Science to them is a strange world, and they are not prepared to come up with any viable solutions. Problems that have been created by the scientists must be solved by the scientists. The social science approach will only further entrench the already polarized positions between the two opposing camps.

What then remains to be done? Simple: *Train scientists who are aware of the problems and prepared to deal with them.*

How is this to be achieved? We have stated that the technology to solve these problems is basically in place. To invent new processes and techniques is part of that existing technology. The particular attribute of *environmental* problems is their *interdisciplinary* nature. The solutions to such problems require the cooperative effort of engineers, geophysicists, engineering geologists, chemists, and biologists. Of these, the only ones who will succeed are those that have a decidedly *broad, technical* background.

While the word environment is certainly a modern buzz word, most people still have only a vague understanding of the actual nature of these problems. It may be useful to provide just a couple of examples at this point.

Alternative energy sources, such as solar and wind energy, enjoy much media coverage. They no doubt have their usefulness in special cases. However, they do not qualify as a general replacement for the fossil fuels (oil, natural gas, coal) for the same reason that the current world population makes absolutely staggering demands on these resources (3 billion tonnes per year oil; 1.7 billion tonnes per year oil equivalent in natural gas; 2 billion tonnes per year oil equivalent in coal[1]).

Most people are unaware of the fact that potable, fresh water occurs only in a thin veneer on our continents. Besides being polluted, it is being severely depleted in many parts of the world (dropping groundwater tables, over allocated stream flows). We have in the past decades not been using the ground— and surface—waters as a renewable resource, but rather have been "mining" them, with a resultant definite end in sight.

To foster a critical awareness of the truly frightening nature of the problems is another requirement of modern technical training. The problem, as I see it, is summarized in statements 1 to 3 in Figure 45.1. The solution is captured in actions 1 to 2.

The one year course might be dubbed a diploma[2] in environmental communication. This addresses the fact that the solution to environmental problems lies in an interdisciplinary approach rather than a concoction of as yet unknown technologies. Some proposed courses are presented in Figure 45.2.

Comments on the one year program in environmental communication:

1. This one year program, following the base degree, should be open to applicants with a degree in engineering, geophysics, engineering geology, biology, chemistry, and possibly additional science degrees.

2. Teachers listing their courses in this program— approved by the coordinator—must provide a 10 page

STATEMENT #1: Environmental problems are ORDINARY technical difficulties of HIGH COMPLEXITY with a definite SOCIAL COMPONENT.

STATEMENT #2: Competent professionals with training in the applied sciences or engineering will POOL their talents towards possible solutions.

STATEMENT #3: No special technical training is required for environmental problems. What must be included in the basic training of applied scientists is the awakening of an AWARENESS and a COMMITMENT for and to the environment.

ACTION #1: Pooling of talents requires a broader background than the current 4-year programs can provide. Offer a 1-year course program consisting of existing electives to eliminate this shortcoming and facilitate the interdisciplinary dialogue.

ACTION #2: Produce three topic oriented courses containing environmental case studies, such as the ramifications of big dams; the depletion and pollution of groundwater; considerations for nuclear waste repositories, etc. (see Figure 45.2).

FIGURE 45.1: A RATIONAL APPROACH TO ENVIRONMENTAL CONCERNS.

#1: ENVIRONMENTAL IMPACT OF OIL/GAS & MINERAL EXPLORATION, EXPLOITATION AND TRANSPORTATION.

#2: ENVIRONMENTAL ASPECTS OF FRESH WATER (SURFACE AND SUBSURFACE) POLLUTION AND DEPLETION.

#3: ENVIRONMENTAL ASPECTS OF REPOSITORIES: FROM HOUSEHOLD GARBAGE TO NUCLEAR WASTE.

#1: Surface disturbance due to seismic lines, trenching, drilling, marine (dynamite charges); disposal of waste rock and fluids; subsidence due to excavation or fluid pressure reduction; pipeline hazards; pelletization of sulphur; industrial emissions.

#2: Chemical pollution of surface and subsurface waters; heating of surface waters; consequences of dam building; consequences of major diversion of surface waters.

#3: Even a simple modern garbage dump is a time bomb. The situation gets worse for the disposal of highly toxic chemicals and in particular low and high level radioactive waste; safety of subsurface gas/oil storage; disposal versus storage; natural hazards such as earthquakes/flash floods.

These courses should be strictly technical without further solidfying already highly polarized positions.

Other topic oriented courses are possible, for example:

#4: THE ENERGY SITUATION: Fossil fuels versus alternative energies; consumption and reserves; global distribution of various energy sources; efficiency of different energy sources; environmental impact of various energy sources.

FIGURE 45.2: SPECIFIC, NO-NONSENSE ENVIRONMENTAL COURSES.

summary of the essence of the required prerequi-
sites, complete with a reading list, as obviously all
students will have shortcomings in some prerequi-
sites. The extra burden of catch-up studies will be
recognized by limiting the required semester hours
to about 30.

Editor's Note
This chapter was written in 1993 as the environmental movement was gaining momentum and the need for a serious
adjustment on the part of the technocrats became increasingly evident.

46

REVOLUTIONS: PAST AND FUTURE

Depending on your temperament, the word "revolution" sends a shudder down your spine or an itch to your trigger finger. In either case, you have the image of rolling heads, blood flowing in the gutter, and bodies floating down the river. Of course, this is not what we are thinking here. Political revolutions have never really changed anything, except personal fortunes. To most of us, revolutions are either devastating—if you have a vested interest in the status quo—or exciting—if you have nothing to lose and everything to gain. But really, all the word designates is that in contrast to evolution, a sudden, drastic, and decisive change has taken place. Since such a change often spells doom for many, the term also carries the connotation of catastrophe and destruction.

We have to tear ourselves loose from these misleading meanings. Throughout his history, man has experienced a number of revolutions that have truly changed his collective existence. The agricultural revolution, some 10,000 years ago, has had a decisive influence on all people, not just some parts of the population. Earlier than that, the conquest of fire, some 500,000 to 1,000,000 years ago, and the invention of distant weapons such as the bow, may have had similar effects. Most recently, the industrial revolution has once more changed the lives of all, or almost all, people. In each case, the previous revolution contains the seeds for the next. The industrial revolution would not have been possible unless preceded by the agricultural revolution.

Figure 46.1 shows the major revolutions in man's history. Needless to say there are many minor peaks on this chart, such as the invention of the wheel, the sail, gunpowder, and others. Again, it bears repeating that most so-called historic events are insignificant and only represent minor eddy currents on the broad river of human history. We have experienced only three true revolutions: the taming of fire, the emergence of the cities, and the current industrial-technological revolution. The latter is completely undigested, regardless of religion, race, or political conviction.

The rate of these revolutions has not been staggering. Hundreds of thousands of years elapsed between the acquisition of fire and the agricultural revolution. Ten thousand years elapsed between the agricultural and industrial revolutions. Not very impressive, but an accelerating trend is clearly indicated. And now the industrial-scientific revolution has produced a state of disequilibrium. Unless it is followed closely by the human revolution (see Chapter 47), mankind cannot survive. Thus, we must have the human revolution in short order—tens of years—to restore equilibrium.

Unless the human revolution can become a reality within the near future, *Homo sapiens* will prove to be nothing more than the latest miscarriage of evolution.

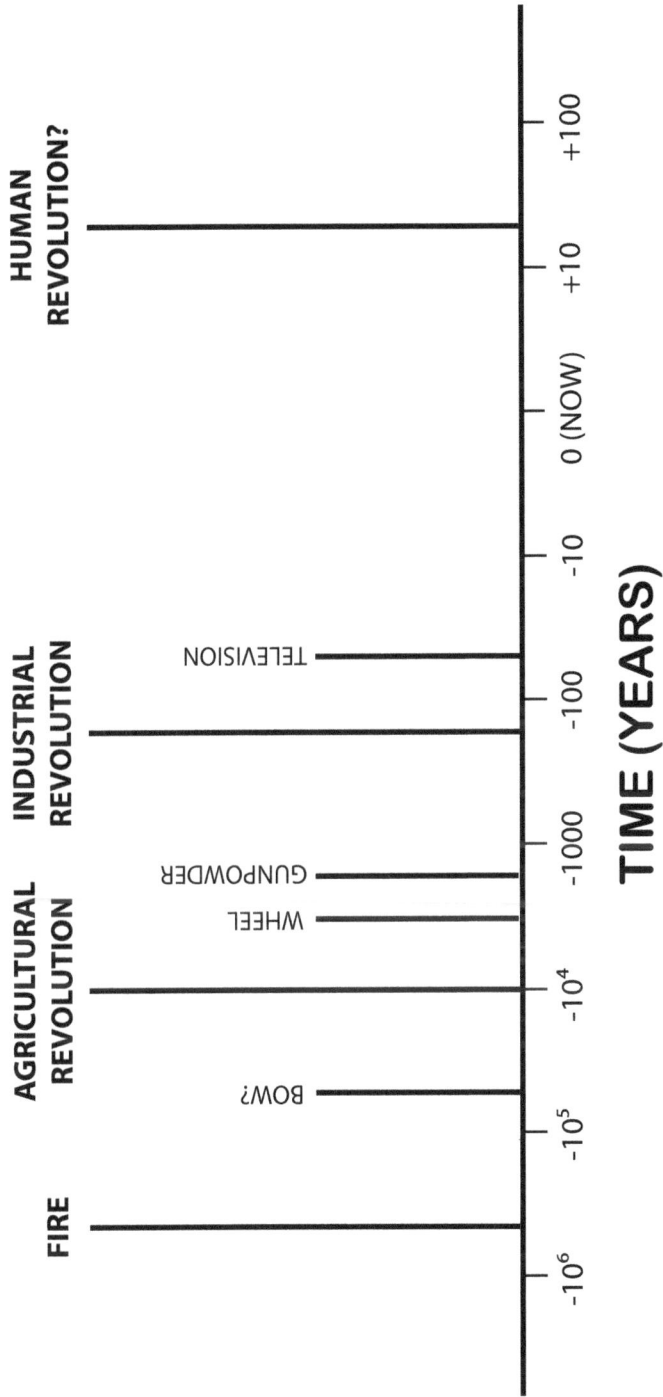

FIGURE 46.1: MAJOR REVOLUTIONS IN MAN'S HISTORY. Major and minor transformational events in man's history. Whether the Human Revolution is achieved remains to be seen.

47

THE HUMAN REVOLUTION

The term "human revolution" refers to a change of attitude in man for which there is no precedent in his recent history. It is called a revolution because any calculations make it clear that it has to occur very quickly in order to be effective.

It is not new, and has been recognized by many, that the solution to modern man's predicament is not found in technology but rather in human attitude.[1] A drastic and sudden change is required. Since we have no evidence that in this respect man has changed at all during the past 10,000 years, it seems appropriate to speak about a revolution.

Also, there seems to be unanimous agreement that time is running out, and such a change has to occur within decades, not centuries. For a cultural change of such a profound nature, this is fast and lends further credibility to the term "revolution."

The nature of such a revolution is fairly clear—it has to make man a more inwardly oriented creature. While we have seen that there is a maximum tolerable level of material affluence (see Chapter 15), there is no such ceiling to be imposed on spiritual and intellectual affluence. Such activities do not produce the harmful side effects associated with material consumption.

The change required is one that turns away from self-centredness and personal greed. The modern predicament as we observe it today is caused by the large human population, but equally by the insatiable human appetite for material goods, leading to an excessive index of affluence. The consumer burden placed on the earth today is largely, or solely, the responsibility of the affluent minority. We not only consume excessively in our working time, but all modern leisure is essentially centred around consuming, such as running powerboats, cars, snowmobiles—excessively replacing recreational equipment by the latest available model or gizmo. Little time is spent on reflective, and thereby non-damaging, activities such as reading, playing chess, and simply relaxing. Hiking, while becoming more popular, does not involve a major portion of the population. Sailing, and in particular rowing, has gone out of style in favour of motor boating. Even the smallest of lawns requires a power mower and any kind of muscle work is avoided. The lack of exercise is then made up with special workouts, often also energy consuming. Packaging has gotten out of hand, again for the sake of convenience. There is a great reservoir of untapped energy—simple manpower. Most westerners never work off their caloric intake by physical work at home.

While the goal is reasonably definable, the path to this goal is far less well laid out. In fact, the present situation is not at all encouraging. The above described changes in man toward inward reflection, contemplation, and more intellectual exercises are not enhanced by either substandard existence or excessive affluence. However, these seem to be exactly the conditions currently prevailing in various parts of the world. The only factor in favour of such a revolution occurring at this time, is the growing awareness of its need or, to put it more realistically, its inevitability, if disaster is to be avoided.

It is further interesting to note that no political system seems to create a more favourable climate for the occurrence of such a revolution. All major systems today are consumption oriented. All of them are tolerant towards ordinary crime, although for different reasons, and all of them are mesmerized by the options offered by modern technology. All, with the possible exception of China, do not encourage the proper steps leading to a solution. It seems that with respect to taming the insatiable material appetites of man, no creed, political system, or religious philosophy has succeeded in coming to grips with this universal and primeval trait of man.

In all western countries, energy consumption could be reduced anywhere up to 30 percent without any effect on the standard of living simply by cutting waste and accepting some inconveniences. To effect this change requires an adjustment in human attitude rather than any kind of technological advance. To stubbornly refuse to even consider such alternatives is to reject the human revolution, as Beckerman does.[2] This can only spell doom, since man may be powerful, but, unfortunately, he is not almighty.

There does seem to be some agreement amongst the optimists and the pessimists. Even Maddox cannot remain an optimist unless he makes tacit assumptions of change.[3] To refute the human revolution places one in the position of Beckerman, who rejects even the most fundamental aspects of the limits to growth and, in essence, maintains that there are *no* limits to growth. Obviously, this is unacceptable.

The short time available makes it clear that such a change cannot come about through any kind of biological evolution. Revolutions of this type can only be achieved through the cultural route. The fact that no progress has been made along these lines despite the attempts of all major religions to tame man's ego and material desires, is not encouraging. However, never before has disobedience had such fatal implications. Not until recently have we had so thorough an understanding of our position on the earth and our relationship to other forms of terrestrial life.

True, much still remains nebulous and uncertain. The nature of our aggression is still a matter of debate. There is little doubt, however, that it is enhanced by crowding, and this is one of the reasons why it is getting worse. We are still not sure to what extent we are a product of our nature and to what extent it is our nurture that is responsible for who we are. But while it is certain that human knowledge will never be perfect, we cannot neglect the fact that today we know more than ever before. The sad aspect, of course, is that so few have any interest in acquiring and enhancing this knowledge. In order to give the human revolution a chance, we must place these problems before the people. Debate and involvement must precede any possible change.

While China still remains an enigma for most of us, there seems to be agreement amongst those knowledgeable that there has been a tremendous change from the old to the new China. In that sense, this example may provide a glimpse of hope that man is, indeed, malleable, and a human revolution is not an impossibility.

The alternative to the human revolution is very simple and straightforward: *Homo sapiens*, the stupid son of a bitch, has had it.

Editor's Note

Harsh? Yes. It reflects the frustration of the author, who sees the promise of the species and, indeed, the species itself, needlessly at risk.

True? Stay tuned. According to the author, it's up to us— all of us.

48

THE COMMANDMENTS OF
THE HUMAN REVOLUTION

1. Thou Shalt Use Your Head

Man is capable of abstract thought, he is able to reflect and so on. Unfortunately, very few do and this amounts to a negation of our own evolution, the evolution of the brain. This is not permissible. The notion that only those who have an appropriate job should do any thinking is erroneous. We are *all* members of the species *Homo sapiens,* and everyone is under the obligation to exercise his or her brain. Many of us today believe that democracy is the form of government that represents the least evil. It is based on the participation of all. Decisions are made that affect all, and as Dürrenmatt said, "What concerns everyone can only be resolved by everyone."[1]

2. Thou Shalt Give Your Fellow Man a Fair Shake

Nothing new. Contained in all major religions. However, in an overcrowded world, more relevant than ever and, unfortunately, less practiced, as the individual merges into the needs of the anonymous mass and feels unobserved.

3. Thou Shalt Not Be a Waste Maker

Waste making is an indefensible crime. Born out of laziness, it contributes nothing to the standard of living and is a direct outcome of the violation of the first Commandment. Note that the first Commandment is in direct conflict with the goals of the consumer society. Any amount of reflection, no matter how minimal, would never have permitted the rise of the consumer society as a viable way of life. It will be difficult to erase.

SELECTED READING RELEVANT TO THE VANISHING OF A SPECIES?

Below are some selected books relevant to *The Vanishing of a Species?* Note that this list is not complete and that comments are purely personal opinions, which you may or may not share. It is left to your resourcefulness to find other, equally, or more relevant, texts.

Vance PACKARD
A good, common sense man who has pointed out many flaws of our society before a more general awareness has set in. Most recommended are his earlier books, such as *Hidden Persuaders, The Waste Makers,* and *The Status Seekers.*

Pierre BERTON
The Comfortable Pew. A critical look at the church in the new age. A study commissioned by the Anglican church. Very thought provoking.

Robert ARDREY
African Genesis, The Territorial Imperative and *The Social Contract.* A powerful and entertaining writer not afraid to draw conclusions. Strongly opposed by many, but good common sense is apparent in his arguments.

Ashley MONTAGU
Editor of *Man and Aggression* – a series of articles intended to refute Ardrey's conclusion that man has been a killer of his own species for a LONG time. The argument whether human aggression is a product of the biological or the cultural evolution of man is addressed. On the whole, this series of articles seems weak to me, but you should not fail to assess the other opinion and draw your own conclusions.

Konrad LORENZ
On Aggression. A well known animal behaviourist and ethicist tells his story. He makes many interesting observations.

Recent Nobel Prize winner. Opposed by some, not least because of his past during Nazi Germany.

John CHRISTOPHER
No Blade of Grass. A major catastrophic event wipes out a large percentage of the earth's population. What Christopher thinks will happen when "the chips are down." If you don't like it, read a similar account by one who has more faith in humanity:

Albert CAMUS
The Plague.

Alexander SOLZHENITSYN
The greatest in my book. *First Circle, The Cancer Ward,* and *One Day in the Life of Ivan Denisovich* are masterpieces. His short Nobel Prize lecture offers the most food for thought per dollar. I believe that it is necessary to undergo the experiences of a Solzhenitsyn in order to develop such perception. However, again just yesterday a student claimed that these experiences simply drove this author around the bend. Make up your own mind.

Gaylord SIMPSON
The Meaning of Evolution. The closing chapter has much to say about man. Simpson is one of the most outstanding minds in the field of geology. He has managed to pack a great deal of wisdom into terse statements. Remember my many quotes.

Björn KURTÉN
Not From the Apes. Another serious account of man's early history.

Donella and Dennis MEADOWS et al.
The Limits to Growth and their antagonist:

John MADDOX
The Doomsday Syndrome. We have discussed this topic sufficiently in the preceding pages. It remains for you to weigh the arguments and form your own opinion.

Thomas Robert MALTHUS
An Essay on the Principle of Population. Written in 1798! A much mentioned classic, available as a Pelican Classic.

SELECTED EXCERPTS FROM
THE VANISHING OF A SPECIES?

M an is a newcomer. He is an integral part of our planet, but not an essential one. There is no natural law that guarantees his survival. (Chapter 3)

Is man a *flash in the pan* in terms of life on earth? (Chapter 4)

It is important to distinguish between the *improbable* and the *impossible*. I postulate the improbable when I claim that a major meteorite will hit the Atlantic within the next five years. I am not likely to be right, but there is no physical law that would preclude such an event. For the tabloid "News" to declare "Baby Born with Wooden Leg" is to advocate the impossible, since wood is not known to grow in the womb. An event with 95% probability of occurrence in the next 15 million years is not a major concern to the human race. However, to relegate it into the realm of the impossible is not permissible for those studying the earth's history. (Chapter 4)

Life has always been risky and will remain so. The joy of being born includes the inevitable acceptance of death. (Chapter 4)

There might be a critical population level beyond which "man simply goes ape," the reverse of what we normally contemplate. Since we evidently do not love each other, to exceed a critical density in packing may be fatal from a social rather than physical point of view. (Chapter 7)

The major religions preaching to give your fellow man a fair shake are correct in terms of man's survival—and yet, several thousand years of preaching have left no mark! (Chapter 7)

If the cultural evolution is unable to produce what on the human timescale would be called a revolution, then one can indeed condense the future of mankind into *"Homo sapiens, the stupid s.o.b., has had it."* (Chapter 8)

We are like visitors, irresponsible ones at that—from outer space—who exploit this place while it lasts, and then move on not caring what, if anything, we leave behind. (Chapter 9)

Our complete dominance is closely related to our success as super toolmakers. This is responsible for our rapidly increasing numbers and our ability to move into every corner of the world, relentlessly pursuing all that creeps, runs, flies, or swims. (Chapter 9)

The term "revolution" evokes in all of us visions of rubble, destruction, decaying bodies, and the accompanying stench. Our thinking is directed to barricades, flags, gunshots, and the final establishment of fairness and justice. The reality, of course, is quite different. In a nutshell: "One son of a bitch takes over from another." And the misery continues. (Chapter 11)

All human systems—political, economic, and religious—have the capacity to function. The fact that none of them does function is not the fault of the systems but rather due to the fact that they all must rely on the same deficient "brick"—*Homo* not so *sapiens*. (Chapter 11)

I have stated, and stand by my guns, that the best translation of *Homo sapiens* is "son of a bitch"—rude, not very flattering, and quite obviously totally unacceptable since it is lacking in scientific rigour. Examine your first reaction: Is it? Human history is an unbroken chain of misery, tears, rubble, and violence—practically all self-inflicted. Contrary to common belief, Mother Earth has been kind to us. Sure, occasionally we get decimated by floods, landslides, and earthquakes. However, that toll is infinitesimal compared to the one we inflict upon ourselves. (Chapter 11)

Greed just makes no sense when all possessions have to be

surrendered in what is, after all, a short time (geologically speaking). (Chapter 11)

A more realistic attitude on our part towards the universe, the earth, and other forms of terrestrial life will go a long way towards solving our problems. (Chapter 11)

A destitute but wise human being is an evolutionary success, but a wealthy and almost all-powerful moron is doomed to extinction. (Chapter 13)

If the term *Homo sapiens* remains the designation for a mechanical genius and a spiritual imbecile, the fate of the species is, indeed, sealed. (Chapter 13)

Where are our leaders? The politicians muddle along according to the wishes of their constituents, the clergy has faded into the background of anonymity, and the academics are busy pursuing their research—read egos—in their ivory towers. (Chapter 16)

To be counted in with the silent majority must be about the worst insult that can be thrown at a thinking human being. In a time when it is painfully obvious that *the* system has had it—when it is equally obvious except to the very youngest that there is no place to run—in such a time to be silent and to be, in fact, an intellectual parasite, is one of the worst crimes a citizen of a free country can commit. It is equal to high treason in times of war. (Chapter 20)

The lack of integrity is somewhat puzzling in a basically affluent society. It is displayed to a large extent by people who have no need for it, i.e. the affluent segment of the population. It is borne out of greed—the insatiable desire for more and more material goods and the need to have everything that others have. It leads to the hectic rat race of modern western nations and usually ends by the racer being buried prematurely, a victim of his addiction. Whatever he has amassed remains behind—at least there are no known cases where someone has taken it with him. (Chapter 21)

Physical laziness is easily visible and demonstrable while intellectual laziness is far more subtle and less obvious, but at the same time its consequences are far more devastating. (Chapter 24)

We come to realize today that in the western world we are presently enjoying a standard of living which we do not deserve; in fact, we have mortgaged the future with our behaviour. (Chapter 31)

The fact that the urge for material possessions is insatiable in man will prove to be his downfall. The invention of property in itself is not bad, what is bad is the fact that there is no ceiling. (Chapter 33)

It is customary today in affluent societies to subscribe to a waste of material goods which is absolutely unnecessary. Mental and spiritual freedom, the most important aspects of a dignified life, can be maintained without the presently associated high level of material consumption. On the contrary, the material consumption, and the energies directed in the production and consumption of these material goods, detracts from enjoying the more leisurely aspects of life. (Chapter 42)

The continuation of present trends spells doom for the western societies. It is these societies that today have the option of disappearing quietly, as others have in the past, or with a bang, endangering the whole of mankind. The possibility that they choose the bang cannot be ruled out. (Chapter 42)

In fighting boredom, we blow away more resources between Friday night and Monday morning than during the entire working week. Conclusion: the enormous resource appetite (and concomitant pollution) of the overdeveloped countries is largely fuelled by our desire for entertainment and is not required for a basic, comfortable existence. (Chapter 43)

Green movements are hell bent on saving the planet. Well, let's just be realistic. The *planet* is doing just fine. True, it has a little skin cancer, deadly to the human race, but hardly of any consequence to the planet. (Chapter 43)

Does it make sense to explore the short span of modern man's history without ever measuring it against the background of the history of our planet, or at least the history of life as we know it today? I submit that such an approach might induce historians to take a somewhat humbler view of our species, much to the advantage of us all. (Chapter 44)

Where man himself is involved, and we look essentially at ourselves, we are apt to become emotional and our unbiased scientific minds tend to be shrouded in the smoke that rises from our burnt egos. (Chapter 44)

History tells us that climate is not a constant. Warm and cold periods alternate. In each case habitable regions are opened and closed, which leads to mass migrations. In an overpopulated world, such migrations are difficult to accommodate. The core problem is not the climate but overpopulation. (Chapter 44)

It is one of the foremost tasks of tenured academics to pound into the western business world that only doom and gloom can result from unrestrained material growth. (Chapter 44)

The interdisciplinary dialogue is *not* a natural activity for humanity. It does not come easily to an ambitious, aggressive, and egocentric animal such as man. It cannot and never will be a widespread and universal activity. In order to bring about a minimum level of interdisciplinary dialogue, we shall have to outwit ourselves. (Chapter 44)

Modern man's predicament is not as obvious as being at war, though the consequences are far more serious. (Chapter 44)

A permissive society in an overcrowded world is an asinine concept. (Chapter 44)

Excellence *next door* is not *excellence* but *competition*. Excellence only acknowledges other excellence when it is safely removed in either space or time, from which it follows that the best excellence is dead. (Chapter 44)

Excellence and egocentricity are, unfortunately, closely linked. Some excellences are constantly surrounded by crackling electricity, and a relaxed conversation—the essence of a successful interdisciplinary dialogue—is just not possible. They must be kept in a special "cage," and the manager assumes the role of zookeeper. This may sound ridiculous to some, but the fact that in most European universities the geosciences, for example, are split into so many departments—much to their detriment—speaks for itself. They learned early on that when you have two male tigers and keep them in the same cage, you soon only have one left. (Chapter 44)

To expect social scientists to find solutions to basically scientific problems is ludicrous. Science to them is a strange world, and they are not prepared to come up with any viable solutions. Problems that have been created by the scientists must be solved by the scientists. (Chapter 45)

Unless the human revolution can become a reality within the near future, *Homo sapiens* will prove to be nothing more than the latest miscarriage of evolution. (Chapter 46)

In all western countries, energy consumption could be reduced anywhere up to 30% without any effect on the standard of living simply by cutting waste and accepting some inconveniences. To effect this requires a change in human attitude rather than any kind of technological advance. To stubbornly refuse to even consider such alternatives means to reject the human revolution. This can only spell doom, since man may be powerful, but, unfortunately, he is not almighty. (Chapter 47)

The alternative to the human revolution is very simple and straightforward: *Homo sapiens*, the stupid son of a bitch, has had it. (Chapter 47)

ACKNOWLEDGEMENTS

My appreciation and thanks: To my friends in Canada, the United States, and Switzerland, who in many discussions over the past few years have helped me formulate the ideas expressed in this book. This is not to imply that they agree with my conclusions. Dissent is more stimulating than agreement.

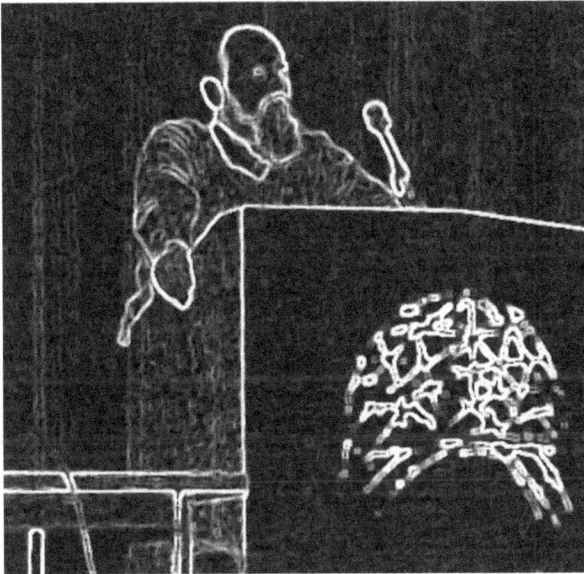

NOTES

FOREWARD

1. Tim Jackson, *Prosperity without Growth? - The Transition to a Sustainable Economy* (United Kingdom: Sustainable Development Commission, 2009), 5 - the report is available at: http://www.sd-commission.org.uk/publications/downloads/prosperity_without_growth_report.pdf.

CHAPTER 1
INTRODUCTION

1. C.P. Snow, *The Two Cultures and A Second Look: An Expanded Version of The Two Cultures and the Scientific Revolution* (Cambridge: Cambridge University Press, 1964).

2. G.K. Gilbert, "The Origin of Hypotheses" (presidential address, Geological Society of Washington, Washington D.C., December 11, 1895).

CHAPTER 3
MAN AND TIME

1. Frank Press and Raymond Siever, *Earth* (San Francisco: W.H. Freeman, 1974), 494.

CHAPTER 4
FURTHER IMPLICATIONS OF LONG TIME SPANS: AWARENESS OF EVOLUTION AND THE RARE EVENT

1. George Gaylord Simpson, *The Meaning of Evolution: A Study of the History of Life and of Its Significance for Man* (New Haven, Connecticut: Yale University Press, 1949), 328-29.

2. Ibid., 301. Note Simpson here is referring to an original quote from what he cites as "...the following thoughtful and valuable paper: C.D. Leake, "Ethicogenisis" (*Sci. Monthly*, 60 [1945], 245-53)."

3. Peter E. Gretener, "Significance of the Rare Event in Geol-
    ogy," *The American Association of Petroleum Geologists
    Bulletin*, Vol. 51, No. 11 (November, 1967): 2197-2206.

4. Values supplied by M. Nosal, Department of Mathemat-
    ics, University of Calgary.

5. Peter E. Gretener, "Continuous versus Discontinuous and
    Self-Terminating versus Self-Perpetuating Processes in
    Geology," *Catastrophist Geology* 211 (1977): 24-34.

6. George Wald, "The Origin of Life," *The Physics and Chem-
    istry of Life* (New York: Simon and Schuster, 1955), 3-26.

7. Norman D. Newell, "Crises in the History of Life," *Scien-
    tific American,* Vol. 208, No. 2 (1963): 76–92.

8. Robert S. Dietz, "Meteorite Impact Suggested by Shat-
    ter Cones in Rock," *Science,* Vol. 131, No. 3416 (June 17,
    1960): 1781-84.

9. Newell, *Crises in Life.*

10. Wilfred Beckerman, *In Defence Of Economic Growth* (Lon-
    don: Cape Press, 1974).

11. Simpson, *Meaning of Evolution,* 292.

12. Pierre Teilhard de Chardin, *The Phenomenon of Man* (New
    York: Fontana Books, 1969), 268.

CHAPTER 5
FROM EARLY MAN THE TOOLMAKER
TO PRESENT HOMO TECHNICUS

1. Richard E. Leakey and Roger Lewin, *Origins: What New
    Discoveries Reveal about the Emergence of Our Species
    and its Possible Future* (New York: E.P. Dutton, 1977), 107.

2. William W. Howells, *Mankind so Far: Man's History, Past,
    Present, and Probable Future* (Garden City, New York:

Doubleday, 1944); *Mankind in the Making: The Story of Human Evolution* (Garden City, New York: Doubleday, 1959); "Cranial variation in man," *Papers of the Peabody Museum of Archaeology and Ethnology*, Vol. 67 (Cambridge, Massachusetts: Harvard University, 1973).

3. Theodosius G. Dobzhansky, "Changing Man," *Science,* Vol. 155, No. 3761 (January 27, 1967): 409-15.

4. Hans Hass, *The Human Animal: The Mystery of Man's Behavior* (New York: Dell Publishing, 1970), 100-09.

5. Ashley Montagu and M. Francis, "On the primate thumb," *American Journal of Physical Anthropology*, Vol. 15, No. 2 (1931): 291-314; Ashley Montagu, "Toolmaking, hunting, and the origin of language," *Annals of the New York Academy of Sciences,* Vol. 280 (October, 1976): 266-74.

6. Jane van Lawick-Goodall, *In the Shadow of Man* (London: Collins, 1971).

CHAPTER 6
HOW UNIQUE IS MAN?

1. Simpson, *Meaning of Evolution*, 281 (see chap. 4, n. 1).

CHAPTER 7
MAN:
A PRODUCT OF HIS NATURE AND HIS NURTURE

1. Theodosius G. Dobzhansky, *Heredity and the Nature of Man* (New York: Harcourt, Brace & World, 1964).

2. John Lewis and Bernard Towers, *Naked Ape or Homo Sapiens?* (New York: Mentor Books, 1973), 69.

3. Ashley Montagu, *The Direction of Human Development* (London: Watts, 1957), 48.

CHAPTER 8
THE QUEST FOR MAN'S NATURE

1. Raymond A. Dart and Dennis Craig, *Adventures with the Missing Link* (New York: Viking, 1959).

2. Robert Ardrey, *African Genesis: A Personal Investigation into the Animal Origins and Nature of Man* (New York: Dell Publishing, 1961), 185-203.

3. Konrad Lorenz, *On Aggression* (London: Methuen & Co, 1966), 205.

4. Ashley Montagu, ed., *Man and Aggression* (London: Oxford University Press, 1968).

5. Valerius Geist, University of Calgary (personal communication).

6. Beckerman, *Defence Of Economic Growth* (see chap. 4, n. 10).

7. Ardrey, *African Genesis,* 150.

8. Lewis and Towers, *Naked Ape,* 90 (see chap 7, n. 2).

9. Ardrey, *African Genesis,* 177-210.

10. Lorenz, *On Aggression,* 205.

11. Robert Ardrey, *African Genesis; The Territorial Imperative: A Personal Inquiry into the Animal Origins of Property and Nations* (New York: Dell Publishing, 1966); and *The Social Contract: A Personal Inquiry into the Evolutionary Sources of Order and Disorder* (New York: Dell Publishing, 1970).

12. Montagu, *Man and Aggression.*

13. Lewis and Towers, *Naked Ape,* 13.

14. See for e.g. Eugène N. Marais, *The Soul of the Ape* (Middlesex England: Penguin Books, 1973).

15. Lorenz, *On Aggression.*

CHAPTER 9
THE MOST DOMINANT SPECIES EVER

1. Simpson, *Meaning of Evolution*, 335 (see chap. 4, n. 1).

CHAPTER 11
THE TWO CULTURES

1. Snow, *Two Cultures* (see chap. 1, n. 1).

2. F.R. Leavis, *Two Cultures? The Significance of C.P. Snow* (New York: Pantheon Books, 1963).

3. The Sir Robert Rede Lecture is an annual public lecture presented at the University of Cambridge. It is named after Sir Robert Rede, who was Chief Justice of the Common Pleas in the sixteenth century.

4. Simpson, *Meaning of Evolution*, 313 (see chap. 4, n. 1).

5. Snow, *Two Cultures*, 6.

6. Leavis, *Two Cultures?*

7. Snow, *Two Cultures*, 23.

8. Ibid., 29-30.

9. Ibid., 40.

10. Georg W.F. Hegel, *Lectures on the Philosophy of History*, trans. J. Sibree (London: G. Bell & Sons, 1905).

11. Snow, *Two Cultures*, 42.

12. Ibid., 49.

13. Leavis, *Two Cultures?*; M. Yudkin, *Sir Charles Snow's Rede Lecture* (New York: Pantheon Books, 1963).

14. Mikhail Agursky, "Contemporary Socioeconomic Systems and their Future Prospects," *From Under the Rubble,* ed. A. Solzhenitsyn (New York: Bantam Books, 1976), 66-86.

15. Simpson, *Meaning of Evolution*, 319.

16. Snow, *Two Cultures*, 50.

17. Ibid., 51.

18. Simpson, *Meaning of Evolution*; Arnold J. Toynbee, *The World and the West* (New York: Oxford University Press, 1953); Peter E. Gretener, "The Interdisciplinary Dialogue – The Challenge of our Times" (lecture, University of Calgary, Calgary, Alberta, 1974).

19. An engineer and personal friend.

20. Teilhard de Chardin, *Phenomenon of Man*, 268 (see chap. 4, n. 12).

21. Snow, *Two Cultures*, 40.

22. Simpson, *Meaning of Evolution*, 325.

23. Lorenz, *On Aggression*, 255 (see chap. 8 n. 3).

24. Ardrey, *Territorial Imperative*, 333 (see chap. 8, n. 11).

25. Toynbee, *World and the West*, 85.

26. Arnold J. Toynbee, *Mankind and Mother Earth: A Narrative History of the World* (New York: Oxford University Press, 1976).

27. Karl-Erik Fichtelius and Sverre Sjölander, *Smarter Than Man? Intelligence in Whales, Dolphins and Humans* (New York: Ballantine Books, 1972), 19.

28. Lorenz, *On Aggression*.

29. Dobzhansky, *Heredity* (see chap. 7, n. 1).

30. See e.g. Montagu, *Man and Aggression* (see chap. 8, n. 4).

31. Jim Al-Khalili, *Black Holes, Wormholes & Time Machines* (London: IOP Publishing, 1999), 89.

32. Stefan Collini, introduction to *The Two Cultures*, by C.P. Snow (Cambridge: Cambridge University Press, 1993), vii.

33. Ibid.

34. John Naughton, "Bridging the Two Cultures" (address, University College Cork, Ireland, September 19, 2002) - a text of the address is available at: http://www.ucc.ie/opa/naughton.htm.

CHAPTER 12
ON THE FRAGMENTATION OF KNOWLEDGE

1. Snow, *Two Cultures* (see chap. 1, n. 1); Toynbee, *World and the West* (see chap. 11, n. 18).

CHAPTER 13
THE CENTRAL PROBLEM:
THE IMBALANCE OF GROWTH

1. Snow, *Two Cultures*, 29-30 (see chap. 1, n. 1).

2. Ardrey, *African Genesis* (see chap. 8, n. 2). In his index under "Tools", Ardrey tellingly says simply "*see* Weapons."

3. Ardrey, *Territorial Imperative,* 333 (see chap. 8, n. 11).

CHAPTER 14
THE LIMITS TO GROWTH REVIEWED

1. Donella H. Meadows et al., *The Limits to Growth* (New York: Universe Books, 1972).

2. Willem L. Oltmans, ed., *On Growth* (New York: Capricorn Books, 1974).

3. $(8 \times 10^{18}$ grains$) \times (10$ cubic centimetres/500 grains$) \times (1$ cubic kilometre/10^{15} cubic centimetres$)$

4. Rome. Food and Agriculture Organization of the United Nations. *FAO Yearbook*, Vol. 25 (1971): 79.

5. M. King Hubbert, "The Energy Resources of the Earth,"

Scientific American, Vol. 225, No. 3 (September, 1971): 70; reprinted as a book (San Francisco: Freeman, 1971).

6. John Christopher, *The Death of Grass* - filmed as *No Blade of Grass* (Middlesex, England: Penguin Books, 1956).

7. John Maddox, *The Doomsday Syndrome* (New York: McGraw-Hill, 1972).

8. Ibid.

9. Beckerman, *Defence of Economic Growth* (see chap. 4, n. 10).

10. Donella H. Meadows, Dennis L. Meadows and Jørgen Randers, *Beyond the Limits: Confronting Global Collapse, Envisioning a Sustainable Future* (White River Junction, Vermont: Chelsea Green, 1993).

11. Matthew R. Simmons, "Revisiting *The Limits to Growth*: Could The Club of Rome Have Been Correct After All" (energy white paper, Houston, October, 2000) - the paper is available at: http://www.simmonsco-intl.com/files/172.pdf.

12. Ibid, 71-72.

13. Donella H. Meadows, Jørgen Randers and Dennis L. Meadows, *Limits to Growth: The 30-Year Update* (White River Junction, Vermont: Chelsea Green, 2004).

14. Ibid, 283-84.

15. Graham Turner, "A Comparison of The Limits to Growth with Thirty Years of Reality" (working paper, Commonwealth Scientific and Industrial Research Organization, *Socio-Economics and the Environment in Discussion CSIRO Working Paper Series 2008-09, ISSN: 1834-5638,* June, 2008) - the paper is available at: http://www.csiro.au/files/files/plje.pdf; see also Jeff Hecht, "Prophesy of economic collapse 'coming true'," *NewScientist online,* November 17, 2008, http://www.newscientist.com/article/dn16058-prophesy-of-economic-collapse-coming-true.html.

16. Ibid., 37-38.

CHAPTER 15
POPULATION, AFFLUENCE AND
THE CONCEPT OF EFFECTIVE POPULATION

1. Earl Cook, "The Flow of Energy in an Industrial Society," *Scientific American,* Vol. 225, No. 3 (September, 1971): 135-44.

2. Leavis, *Two Cultures?* (see chap. 11, n. 2).

3. Paul E. Erdman, *The Crash of '79* (New York: Pocket Books, 1977).

CHAPTER 19
ON OPTIMISTS AND PESSIMISTS

1. Albert E. Engel, "Time and the Earth," *American Scientist,* Vol. 57, No. 4 (1969): 460.

CHAPTER 20
THE SILENT MAJORITY

1. Friedrich Dürrenmatt, *The Physicists* (New York: Grove Press, 1962), Appendix "21 Points to the Physicists," Points 17 and 18.

2. This theme resonates in much of Solzhenitsyn's work. In particular, see: Alexander Solzhenitsyn, *Nobel Lecture* (New York: Farrar, Straus and Giroux, 1972); "As Breathing and Consciousness Return," *From Under the Rubble,* ed. A. Solzhenitsyn (New York: Bantam Books, 1976), 1-23.

CHAPTER 21
ON THE LACK OF INTEGRITY AND COMPETENCE

1. Vance Packard, *The Waste Makers* (New York: Pocket Books, 1960).

2. Peter J. Henning, "Another View: The Madoff Scheme," *DealBook online*, December 15, 2008, http://dealbook. blogs.nytimes.com/2008/12/15/another-view-the-madoff-scheme/.

3. Edmund L. Andrews and Peter Baker, "A.I.G. Planning Huge Bonuses After $170 Billion Bailout," *New York Times online*, March 14, 2009, http://www.nytimes. com/2009/03/15/business/15AIG.html.

4. Ron Chernow, "Where Is Our Ferdinand Pecora?," *New York Times online*, January 6, 2009, http://www.nytimes. com/2009/01/06/opinion/06chernow.html.

5. Ferdinand Pecora, *Wall Street Under Oath: The Story of our Modern Money Changers* (New York: Augustus M. Kelley, 1968), xi.

6. Martin McLaughlin, "Clinton, Republicans agree to deregulation of US financial system," *World Socialist Web Site*, November 1, 1999, http://www.wsws.org/articles/1999/nov1999/bank-n01.shtml.

7. Jonathan Weisman, "Obama Keen to Regulate Finance: Early Initiative Aims to Consolidate Watchdog Agencies and Plug Holes in Oversight," *Wallstreet Journal online*, December 19, 2008, http://online.wsj.com/article/SB122965186108420649.html.

CHAPTER 23
COMMUNISM AND CAPITALISM EQUALLY DOOMED

1. Arnold J. Toynbee in *On Growth*, 31 (see chap. 14, n. 2).

2. Lorenz, *On Aggression*, 250 (see chap. 8, n. 3).

CHAPTER 25
CHINA: A CLOUD OR A RAY OF HOPE?

1. Mao Tse-Tung, *Quotations from Chairman Mao Tse-Tung* (Peking: Foreign Language Press, 1967), 204-05.

2. Harry Schwartz, *China* (New York: Antheneum, 1965).

CHAPTER 26
SWITZERLAND AND NORTH AMERICA

1. Dürrenmatt, *Physicists* (see chap. 20, n. 1).
2. A Swiss writer and personal friend.

CHAPTER 29
WHAT MAKES A UNIVERSITY:
DIPLOMA OR BACHELOR?

1. Peter E. Gretener, "The Last Great Forum: What Makes a University? - Diploma or Bachelor?," *The Gauntlet - University of Calgary Student Newspaper*, Vol. 16, No. 40 (March 12, 1976): 6-7.

CHAPTER 30
THE WORKING MOTHER: A MYTH

1. Simpson, *Meaning of Evolution,* 319 (see chap. 4, n. 1).
2. "'Raising children properly' requires stay-at-home parent: Alberta minister," *CBC News online*, June 17, 2009, http://www.cbc.ca/canada/calgary/story/2009/06/17/education-iris-evans-alberta-minister.html.
3. John Rodgers, "Octuplet mom had 6 embryos implanted, far more than usual," *Pantagraph.com*, February 6, 2009, http://www.pantagraph.com/articles/2009/02/06/news/doc498c2ff94ef08144181035.txt; "Octuplet's mom: 'All I ever wanted,'" *CNN online*, February 6, 2009, http://us.cnn.com/2009/US/02/06/octuplets.mom/index.html.
4. Suzanne Steel, "Medical ethics questioned in octuplets birth," *National Post online*, January 30, 2009, http://network.nationalpost.com/np/blogs/posted/archive/2009/01/30/Medical-ethics-questioned-in-octuplets-birth.aspx.

5. Matthew Coutts, "Calgarian has twins at 60: Mother's advanced age raises ethical concerns," *National Post*, February 5, 2009.

6. Dave Dormer, "60-year-old gives birth to twins," *St. Catherines Standard (feed from Sun Media) online*, February 6, 2009, http://www.stcatharinesstandard.ca/.

7. Marni Ko, "When Mommy is a Grandmommy," *backofthebook.ca – Canada's online magazine*, June 3, 2007, http://backofthebook.ca/living/2007_06_01_archive.html.

8. "60 Year Old Mother of Twins Defends Choice to Have Kids," *Guide to Egg Donation and Infertility online blog*, http://egg-donation-directory.blogspot.com/2007/05/60-year-old-mother-of-twins-defends.html.

CHAPTER 33
RIP-OFF: A SYNDROME OF AFFLUENCE ANARCHY

1. Vance Packard, *The Hidden Persuaders* (New York: David McKay Company, 1957).

CHAPTER 34
REFLECTIONS ON THE GENERATION GAP

1. Konrad Lorenz, *Civilized Man's Eight Deadly Sins* (London: Methuen & Co., 1974), 46-57.

2. Ibid.

CHAPTER 38
THE CONCEPT OF GOOD AND BAD

1. Simpson, *Meaning of Evolution*, 300 (see chap. 4, n. 1).

2. See generally the works of Alexander Solzhenitsyn – some references appear at page 232.

CHAPTER 39
OUR GREATEST ENEMY: OUR EGO

1. Ardrey, *Territorial Imperative*, 320-53 (see chap. 8, n. 11).

CHAPTER 40
QUO VADIS HOMO SAPIENS?
FUTURE: AN ATTEMPT AT A DEFINITION

1. Simpson, *Meaning of Evolution*, 325 (see chap. 4, n. 1).

2. Ibid., 309-10.

CHAPTER 42
TO STRIKE A BALANCE BETWEEN DOOMSDAY
PESSIMISM AND IRRESPONSIBLE OPTIMISM

1. Meadows, *Limits to Growth* (see chap. 14, n. 1).

2. Maddox, *Doomsday Syndrome* (see chap. 14, n. 7).

3. Beckerman, *Defence Of Economic Growth* (see chap. 4, n. 10).

CHAPTER 43
THE 21ST CENTURY: A GEOLOGIST'S PERSPECTIVE

1. Ardrey, *Territorial Imperative*, 333 (see chap. 8, n. 11).

2. Hubbert, "Energy Resources" (see chap. 14, n. 5).

3. John D. Edwards, "Crude Oil and Alternate Energy Production Forecasts for the Twenty-First Century: The End of the Hydrocarbon Era," *The American Association of Petroleum Geologists Bulletin,* Vol. 81, No. 8 (August, 1997): 1293.

CHAPTER 44
THE INTERDISCIPLINARY DIALOGUE:
THE CHALLENGE OF OUR TIMES

1. Gretener, "Interdisciplinary Dialogue" (see chap. 11, n. 18).

2. Ardrey, *African Genesis*, 207 (see chap. 8, n. 2).

3. Alexander Solzhenitsyn, *Nobel Lecture*, 27 (see chap. 20, n. 2).

4. Simpson, *Meaning of Evolution*, 313 (see chap. 4, n. 1).

5. Maddox, *Doomsday Syndrome* (see chap. 14, n. 7).

6. Engel, *Time and Earth* (see chap. 19, n. 1).

7. George Gaylord Simpson, "The Biological Nature of Man," *Science*, Vol. 152, No. 3721 (April 22, 1966): 478.

8. Bill Fisher, University of Texas (personal communication).

9. Dürrenmatt, *Physicists* (see chap. 20, n. 1).

10. B.F. Skinner, *Beyond Freedom and Dignity* (New York: Alfred A. Knopf, 1971), 102.

11. Simpson, *Meaning of Evolution*, 319.

12. Lynn White, Jr., L., 1967, "The Historical Roots of our Ecological Crisis", *Science*, Vol. 155, No. 3767 (March 10, 1967): 1207.

13. Hegel, *Lectures* (see chap. 11, n. 10).

14. Barack Obama, *The Audacity of Hope: Thoughts on Reclaiming the American Dream* (New York: Crown, 2006).

CHAPTER 45
STEWARDSHIP OF THE EARTH:
HOW TO DEAL WITH
ENVIRONMENTAL CHALLENGES
IN A SENSIBLE AND RESPONSIBLE MANNER

1. BP Statistical Review of World Energy, 1991. Editor's note: These would be numbers for the previous reporting year of 1990. Numbers for 2008 contained in the 2009 Review: 3.9 billion tonnes per year oil; 2.7 billion tonnes per year oil equivalent in natural gas; and 3.3 billion tonnes per year oil equivalent in coal - the BP 2009 Sta-

tistical Review of World Energy is available at: http://
www.bp.com/liveassets/bp_internet/globalbp/globalbp_
uk_english/reports_and_publications/statistical_energy_
review_2008/STAGING/local_assets/2009_downloads/
statistical_review_of_world_energy_full_report_2009.pdf.

2. Any suitable existing degree designation may be substituted.

CHAPTER 47
THE HUMAN REVOLUTION

1. Meadows, *Limits to Growth* (see chap. 14, n. 1); Meadows, *30-Year Update* (see chap. 14, n. 13); Toynbee, *World and the West* (see chap. 11, n. 18); Snow, *Two Cultures* (see chap. 1, n. 1); see also the works of Aldous Huxley and others.

2. Beckerman, *Defence Of Economic Growth* (see chap. 4, n. 10).

3. Maddox, *Doomsday Syndrome* (see chap. 14, n. 7).

CHAPTER 48
THE COMMANDMENTS OF
THE HUMAN REVOLUTION

1. Dürrenmatt, *Physicists* (see chap. 20, n. 1).

INDEX

Symbols

www.ingramcontent.com/pod-product-compliance
Lightning Source LLC
Chambersburg PA
CBHW030800150426
42813CB00068B/3296/J